Lucien Marcus Underwood

Moulds, Mildews, and Mushrooms

A Guide to the Systematic Study of the Fungi and Mycetozoa and their Literature

Lucien Marcus Underwood

Moulds, Mildews, and Mushrooms
A Guide to the Systematic Study of the Fungi and Mycetozoa and their Literature

ISBN/EAN: 9783337205720

Printed in Europe, USA, Canada, Australia, Japan

Cover: Foto ©berggeist007 / pixelio.de

More available books at **www.hansebooks.com**

MOULDS

MILDEWS AND MUSHROOMS

*A GUIDE TO THE SYSTEMATIC STUDY OF THE FUNGI
AND MYCETOZOA AND THEIR LITERATURE*

BY

LUCIEN MARCUS UNDERWOOD

Professor of Botany, Columbia University

NEW YORK
HENRY HOLT AND COMPANY
1899

THE NEW ERA PRINTING COMPANY
LANCASTER, PA.

PREFACE

The increasing interest that has been developed in fungi during the past few years, together with the fact that there is no guide written in the English language to the modern classification of the group and its extensive but scattered literature, has led the writer to prepare this introduction for the use of those who wish to know something of this interesting series of plants.

With nearly a thousand genera of fungi represented in our country alone, it was manifestly impossible to include them all in a pocket guide. A line must be drawn somewhere, and it was decided to include : (1) Conspicuous fleshy and woody fungi, (2) The cup-fungi, since so little literature treating of American forms was available, and (3) Genera containing parasitic species. Most of the genera of the so-called *Pyrenomycetes* and many of the saprophytic *fungi imperfecti* are therefore omitted from special consideration.

It is hoped that for the groups treated, the synopses will be sufficiently simple to enable the average student to distinguish generically the ordinary fungi that he is likely to find. In every order, references to the leading systematic literature have been freely given, in the hope that some will be encouraged to take up the systematic study of some group and pursue it as exhaustively as possible. With all the diversity of interesting lines of research that are constantly opening before the student of botany of to-day, there is none more inviting to a student, or better adapted to bring into activity all the resources of his judgment, than the systematic study of the species of some limited group, provided this is properly combined with a study of the morphology, development, and ecologic relations of such a related series. With very few exceptions, there is no group of fungi that is not in crying need of thorough and original systematic study.

The attempt has also been made in the following pages to meet the popular interest in fungi as an article of food, by treating the fleshy forms with a greater degree of fulness than others, and one should be able from this treatment to discriminate the ordinary edible, suspicious, and poisonous species, and recognize with sufficient certainty what forms are safe to test for food. Those whose interest centers in edible species alone, will find the groups that interest them on pages 63-66, 97-129, and 136-145. It is suggested that, in the field of exploration for edible species, it is safest to make haste slowly, and the novice is hereby warned of the danger of eating any species which is not thoroughly known.

In the general arrangement of the system the writer has largely followed the treatment in *Die natürlichen Pflanzenfamilien*, tho deviations from the sequence of groups there adopted will frequently appear, and group names—orders and families—are made rigidly to conform to the system proposed at Berlin, but indifferently followed in their recent publications.

The writer is deeply indebted to his friend and former colleague, Professor F. S. Earle, for his kindness in reading those parts of the proof relating especially to parasitic species, and for making many valuable suggestions; and to his assistant, Dr. Marshall A. Howe, for much kindly assistance and many useful suggestions.

The colored plate which introduces the book was painted from nature by a former student, Miss Julia E. Clearwaters, and has been faithfully reproduced in six colors by the Heliotype Printing Company of Boston. The nine plates which conclude the work were drawn under the writer's direction by Miss M. E. Baker; these are from various sources which are duly credited in the explanations of figures; a few were drawn direct from nature.

COLUMBIA UNIVERSITY,
   20 July, 1899.

CONTENTS

MORCHELLA ESCULENTA—FRONTISPIECE

Chapter	I.	Introductory,	1
Chapter	II.	The Relations of Fungi to other Plants,	8
Chapter	III.	Reproduction, Constituents, and Habits,	13
Chapter	IV.	Class I. PHYCOMYCETES,	22
Chapter	V.	Class II. ASCOMYCETES,	34
Chapter	VI.	The Fungi Imperfecti,	68
Chapter	VII.	The lower BASIDIOMYCETES,	80
Chapter	VIII.	The higher BASIDIOMYCETES,	94
Chapter	IX.	Fungus Allies—the MYXOMYCETES,	146
Chapter	X.	The Study of Mycology in general and its Study in America in particular,	155
Chapter	XI.	The Geographic Distribution of American Fungi,	165
Chapter	XII.	Methods of Collection and Preservation of Fungi—Hints for further Study,	201

INDEXES

    I. Index to Latin Names, . 209
   II. Index to Host Plants, . 216
  III. Index of Authors and Collectors, . 218
  IV. General Index and Explanation of Terms, . 221

PLATES 1–9 WITH EXPLANATIONS, . 228

MOULDS, MILDEWS AND MUSHROOMS

CHAPTER I

INTRODUCTORY

The world is full of surprises on every hand. To one whose familiarity with plants is limited to the trees and shrubs of parks and groves, or the herbaceous plants of indoor cultivation, or the grasses of limited plots and lawns, or even to one whose walks more happily include the fields and woodlands, it may seem perplexing, perhaps, to be told that the green slimes with which Nature paints the shaded walls and tree-trunks, or that float as a green scum on the surface of pools, or that cover the pots and benches in green-houses, are likewise plants, each in its simpler, less assuming manner carrying on the same functions as the more conspicuous trees and shrubs. The surprise may be still greater when he learns that the gray-green lichens on fences and rocks, the toadstools springing from the ground or old tree-trunks, the puff-balls clustered on old logs or the larger ones growing singly in pastures are also plants. After this he will be more able to believe that the moulds that grow on cheese or preserves, the mildews and blights that spread over cultivated plants to their injury, the smut of corn and oats, the rust of wheat and other cereals, are all likewise plants, each with its own peculiar life history, each with its peculiar method of reproduction, each occupying its definite place in the economy of Nature. And probably the surprise will be greatest of all to those fortunate persons who have not learned to depend on the baker alone for their supply of the staff of life and to whom the process of bread-making is not an obsolete feature of household work, to be gravely informed that the very yeast by which their flour and water is made to rise into the porous spongy dough is just as truly a plant as is the geranium growing at the kitchen window or the maple that shades the kitchen porch, and that the entire process of bread-raising is due to the growth, development and rapid reproduction of a plant quickened into activity by the presence of moisture and a suitable degree of warmth. Surprise may thence degenerate into a shock when persons, even those of the most refined habits, come to learn that in their own persons they support varied and interesting colonies of extremely minute plants which find perhaps the most suitable conditions for their development and multiplication among the papillae of the tongue and about the crowns of the teeth.

Somehow many people associate life with locomotion and while they think of animals as alive, they look upon plants as dead, and upon botany as the study of the dead rather than of the living. There can be no greater mistake, for plants equally with animals are not only thoroughly alive, but from the greater simplicity of their structure offer even better facilities for the working out of problems connected with general biology, the science not of animals alone, but of all living things.

When a compound microscope becomes as much of a household necessity as a clock or a piano; when children are early taught the nature study of every-day life, and become familiar with the common things in nature around them, these ideas as to what the term plant life includes will not only cease to strike us as mysterious, but our range of available information will be infinitely extended. There is no reason whatever why a compound microscope of low magnifying power should not be just as much a common appurtenance of a well-regulated household as a piano or a music-box. Not as an instrument to be kept under a glass case to show to strangers, not as an elaborate piece of mechanism liable to become disarranged by use, but a simple apparatus suitable to be used by intelligent children and an every-day source of instruction and enjoyment.

In our early childhood many of us acquire certain bits of information, too often as the direct result of teaching, that in after life we find ourselves forced to unlearn. Some of these are principles that the books conspired to impress upon us. The dogma that the interior of the earth is molten and that the exterior crust is thinner proportionately than an egg-shell was stated to us in the geographies with all the gravity of established truth, and yet it is supported by the most flimsy evidence and no one at present thinks of it seriously as a possible theory even. We were told that there were just five races of man, and the impression was left upon us that these were sharply defined. We now know that this was merely Blumenbach's classification of a series of closely intergrading types with every shade of color from the blackest Australian to the whitest Caucasian, and that any one of a dozen other classifications of race based on wider data may be vastly more rational. Another unsupported dogma that we learned was that there are three kingdoms of nature, the animal, the vegetable, and the mineral, and thus that a fundamental difference existed between animals and plants. All this we are obliged to unlearn and this "science falsely so-called" of our early youth must retreat before the light of modern investigation.

The establishment of the identity between what had been known as animal sarcode and vegetable protoplasm, followed by the propositions of Darwin and Wallace relative to the origin of species gave an impetus to the study of living things which has resulted in the building up of our present knowledge. The increased use of the microscope in the study of minute forms of life, and the discovery of connecting types has driven us to the conclusion that there is fundamentally no definable difference between animals and plants, and it is useless to attempt to manufacture distinctions where they do not exist. There is one kingdom of living things, and we have come to look upon plant and animal characters as simply marking tendencies which on either hand become more pronounced as we pass from simple one celled organisms towards those that are more highly specialized. No lines drawn between the series of plant and animal forms will serve as a permanent and satisfactory boundary line ; some forms or some stages of development in the life history of forms will be sure to overleap any such artificial boundaries, for they do not exist in nature.

If we believe, as we surely must, that the various higher forms of living things have been derived from lower forms, that new creations, in other words, are simply new births or growths from earlier and simpler forms and so on back to the simplest organisms that first appeared, then we should expect to find in nature connecting types between one species and another, one family and another, and even one of the greater groups and others more or less distinctly separated. This is exactly the condition that we find to exist, and as we descend from the higher and more specialized forms to those that are lower and simpler, we find these connecting links more and more pronounced and connecting larger and larger groups of organisms. Classification instead of being simply a means of separating forms, has become a method of studying affinities, and tracing the phylogenies or race histories of groups of organisms throughout their complicated alliances.

With the understanding then that no hard and fast lines can be drawn between any of the greater groups of living things, we will endeavor to show the three lines of evolution which have been at work and have given us three types of organisms each with its peculiar tendencies leading it toward a different goal.

Whether an organism be high or low, there are certain common principles involved in its existence. In the first place it finds itself face to face with a tremendous struggle for existence, and secondly it puts forth a powerful effort not only to maintain its life, but also to perpetuate its kind. A single garden weed produces a thousand or even ten thousand seeds, and yet most weeds are not prominently increasing in the total number of individuals from year to year; a plant of the rare hart's tongue fern produces perhaps 50,000,000 spores in a season and yet is barely holding its own in the struggle for existence, and in our own country is found only in a few favored ledges of rock mostly of a certain geologic period; the spores of the giant puff-ball number untold millions and each microscopic spore under favoring conditions would be able to produce a new plant, but the giant puff-ball is far from being common. When we consider the fact that each plant in its lifetime has before it the necessity to produce only one seed or spore that shall come to maturity, in order that the plant may hold its own, and when we consider, moreover, that more plants are growing scarcer than are increasing in number, we can comprehend something of the terrible struggle for existence that is everywhere and always in progress and faces every living thing from monad to man, from the instant it appears on the face of the earth.

To perpetuate its kind, then, is the first great instinct of a living thing and methods of reproduction are the functions developed simultaneously with assimilation and growth. Among all the simplest forms of life whose structure is largely confined within the limits of a single cell, asexual reproduction or multiplication by division is the common rule, while the idea of sex was more fully developed among slightly higher forms as the result of new and distinctly higher necessities in the struggle for life.

The distinctions between the simple forms of green plants and animals are physiological rather than structural, since both are simple masses of protoplasm, enveloped or not in a cell wall as the case may be. But the possession of chlorophyl by the green plant renders it a peculiar organism with the ability to utilize the strictly inorganic components of air, water and mineral salts, and through the energy of sunlight manufacture them into complex organic compounds. The chemical function of green plants is thus synthetic, producing complex molecules from much more simple ones.

Animals, on the contrary, depend for their life on organic food, first manufactured by the green plants. This simple difference of function in the simple one-celled organisms has resulted in the important distinctions that rapidly appeared as the one-celled organisms increased in complexity and became the higher animals and higher plants. In the struggle for food, while the plants found their supply in the air and water in which they were bathed, the animals were forced to *seek* their supply and hence arose the principle of locomotion and with this as complexity of structure increased, there came the necessity for an elaborate but compact digestive, circulatory and respiratory system ; with locomotion also came the necessity in the struggle for life to seek safety and avoid danger, hence the elaborate system of special senses that are the peculiar endowment of all the higher animals.

On the other hand, since the plant had no necessity to seek its food, it had no need for locomotion, and except in the case of a few special organs it has never developed the power. Being stationary, compactness cut no figure in its needs, and in proportion as its size increased, it spread itself out in root and leaf to offer as wide an absorptive surface as possible to the media from which it draws its sustenance. Being stationary it has had no need of sight or hearing or the other special senses of animals, because it could neither avoid danger nor retreat to a place of safety had they been developed. On the contrary, it has developed special senses of its own, none the less remarkable than those of the animal and far better adapted to its own peculiar development. It has, for instance, evolved a marvelous relation to gravity and a peculiar sensitiveness to light unknown among animal life, besides other special senses.

Both animals and green plants have developed and maintained sex reproduction. Moreover the animal has developed sex individuality to a very marked degree. Among the plants, a high degree of complexity became possible, only when the principle of alternation of generations reduced the sexual growth to a minimum, and correspondingly magnified the possibilities of asexual development. Although the animal lost its physical independence when it ceased to produce its own food from inorganic matter, by persistently maintaining the idea of sex and sexual individuality, and perfecting its method of locomotion, and with this the acuteness of its special senses, it has made possible the later evolution of the highest possible development of life, refined and perfected in ourselves.

We have thus emphasized the distinctions between green plants and animals because there has been a third line of evolution leading from the simpler green plants which commenced its divergence in nearly the same points as the animals, but which has had a far different history and has attained a widely different development.

Whether we are aware of their existence or not, there are in the world about us a vast array of more or less inconspicuous organisms that are known to botanists under the name of fungi. These differ among themselves in size and structure far more widely than do a violet and an oak, and many of them at first sight would seem to bear so little resemblance to one another as to possess no real relationship. Many of them are known more or less popularly under common names such as moulds, mildews, mushrooms, toadstools, puff-balls, rusts, smuts, leaf-spots, blights—each popular name indicating a more or less indefinite group of plants, more or less closely related to one another. They grow in every conceivable place wherever organic matter can be found which will serve as their food, and a moderate degree of heat and moisture are present to furnish the necessary conditions of growth. Decaying fruits or vegetables, oily bones, old musty shoes, wet paper, the dead stems of herbaceous or semi-woody plants, the dead or dying branches of trees, standing stumps and tree trunks and fallen logs all furnish the matrix in which fungi of various sorts, a few conspicuous, many more inconspicuous, thrive and multiply. With all their differences fungi agree in two characters, one positive and the other negative, that will enable them to be recognized even by the novice : (1) They possess none of the green coloring matter of ordinary vegetation, and (2) They reproduce by spores. The latter character will distinguish them from the few orchids, broom-rapes, dodders and other seed plants which share with fungi the absence of chlorophyl, while the first character separates them physiologically from ordinary green plants as widely as the latter are separated from the animals, since without chlorophyl they have lost their power to live on the constituents of air and water.

Fungi, like animals, early lost this power to appropriate inorganic food, and thus, like animals, became dependent on organic matter for their very life, but unlike animals they (1) Never developed the power of locomotion, and (2) Soon lost the power of sex reproduction which they had developed in common with all living forms. Instead of the high development reached by animals, the fungi have ever remained either scavengers or parasites, and fulfill a lowly and even degraded calling, at times neutral, at times dangerous, but in many cases beneficial to the world of life at large.

We have then two differentiations from the lowest life, alike in many particulars since they are both dependent on green plants for food and are both *destructive* instead of *constructive* chemical agents—the animal rising step by step to the highest scale of organic and sentient being ; the fungus delighting in decay, degraded and destined to the humblest plane of existence.

CHAPTER II

THE RELATIONS OF FUNGI TO OTHER PLANTS

The whole array of plants known to science, of which there are some two hundred thousand different species, may be conveniently grouped in three great divisions, each of which can be fairly well characterized by certain more or less easily recognized marks. The common herbs, shrubs and trees that constitute vegetation in the ordinary sense, each produce some structure recognizable as a flower and develop seeds in their process of generation. Such plants form a convenient and easily-recognized group known as seed-producing plants or *spermaphytes*.

The second great division is not characterized by common structures so well known as seeds, but may be recognized as containing plants which do not develop seeds but still possess a leafy axis. The higher forms with woody tissues are well known as the ferns and their allies, and the lower and simpler forms are known as mosses. From the common form of the egg apparatus found in all these plants they have been called collectively *archegoniates*. Passing still lower down the scale of plant existence we find a series of plants of still simpler structures which do not develop a leafy axis, but which are formed of masses of cellular tissue ranging in various species from a flat leaf-like expansion of cells, through thread-like forms, to series of single microscopic cells entirely separate from each other. Such an irregular expansion of vegetation we call a thallus, and it has become convenient to speak of all such plants as *thallophytes*. Here the green slimes, the pond-scums, the sea lettuce, the rock-weeds, the red algæ and all the hosts of fungi from yeast to mushrooms are included, altho the plants thus associated form a most unwieldly and heterogeneous assemblage of organisms.

Summarized and arranged in reverse order so as to group the simpler forms first, we have the following tabular survey of the vegetable kingdom :

I. THALLOPHYTA	ALGÆ (pond-scums, diatoms, seaweeds). FUNGI (moulds, mildews, mushrooms).
II. ARCHEGONIATA	BRYOPHYTA (liverworts, mosses and their allies). PTERIDOPHYTA (ferns and their allies).
III. SPERMAPHYTA	GYMNOSPERMAE (pines, junipers, yews, etc.). MONOCOTYLEDONAE (grasses, lilies, palms, etc.). DICOTYLEDONAE (Oaks, roses, clover, sunflowers, etc.).

This simple table will aid us in fixing the place of the fungi in the plant world.

The first character to be noted in all these plants, great or small, high in organization or simple in structure, is the fact that they *breathe*. In order for an organism to live it must first breathe, and we have said that like animals, plants are living things. Furthermore, they breathe for exactly the same purpose as animals and their breathing has the same effect both on the air that is taken in, and on themselves. The oxygen of the air is taken up by the plant the same as we take it into our system, and once in the system it combines with the carbon of the plant structure and reappears as carbonic oxide, just as the oxygen in our arterial blood combines with the carbon of our tissues and is thrown out in the air likewise as carbonic oxide. The process of respiration in other words, is an identical process in both plants and animals. We emphasize this point because it is a stumbling block in the way of a correct understanding of the relations of living things to each other, and much of the popular teaching of the day still continues to present the mistaken idea that the respiration process is exactly opposite in plants and animals.

Plants like animals must also have food to renew their tissues and provide for waste and growth. The animal depends for his food on organic substances already prepared for him. No animal is capable of living on purely mineral food, and we have seen, on the other hand, that in addition to their respiration, green plants—an oak, a rose, a moss, and even the microscopic green algae--have the power of decomposing the carbonic oxide of the air, and, with the elements of water added, actually manufacture organic food. As we have seen, this is the common function of green plants and at once distinguishes them from animals on the one side, and fungi and other colorless plants on the other. Fungi, like animals, require organized food on which to live and they must, therefore, like animals, depend on green plants to manufacture for them organic food from the constituents of air and water and mineral salts. The reasons why fungi are not animals, are the very important ones, that in structure, in the chemical composition of their cell-walls, and in their methods of reproduction the fungi are closely related to the algae.

Some 35,000 or 40,000 species are known at present of which perhaps 8,000 are known to inhabit North America. They vary in size from single microscopic cells to systems of entangled threads many feet in extent which develop reproductive bodies as large as a man's head, or even larger. In color they vary from white through yellow, blue and red to black. Although some of them are green they never possess chlorophyl, and this one negative character is their chief distinguishing characteristic.

The species of fungi, small as many of them are, are usually well characterized, at least as well as the species of the higher plants. Some of the difficulties that are experienced in recognizing our species does not arise so much from the lack of inherent differences, but comes from the fact that there is much similarity between the fungous flora of Europe and America and many of their species and some of ours were poorly defined at the beginning, and many of the descriptions of the early writers are difficult to interpret, and to assign to existing forms. There are other difficulties of an entirely different character to which we will allude later. We are sometimes accustomed to group fungi as parasitic when they draw their sustenance from other living organisms, and saprophytic when they live on decaying matter, but there are various grades intermediate between these artificial groups. It will be convenient for our purpose, however, to refer to them in this way.

The various forms of saprophytic fungi grow on all sorts of dead and decomposing matter; the various forms of mould on bread, on cheese, on chestnuts and on fruit are more or less familiar objects to every one and yet few stop to think of them as genuine plants, each with its peculiar mode of perpetuating its kind, each with tastes and habits peculiar to itself, and each as distinct from other forms of mould as pines and hemlocks are from oaks and maples. More or less familiar also are the more conspicuous forms popularly known as toad-stools or puff-balls, or forming inverted brackets on the stumps and half dead trunks of standing trees. Less known are the myriad minute forms found on twigs and branches and dead herbaceous stems and presenting as diverse forms of structure as do the better known and more familiar flowering plants.

When we pass to the parasitic forms of fungi, we find scarcely any limit to their habitat. From present indications it is probable that nearly every species of the higher plants has growing upon it some fungus species that has adopted it as a host and lives at its expense. Some of the higher plants have three or four parasites preying upon them. The common spring anemone, for example, besides fighting its way in the struggle for existence with other plants of its own size and larger, supports a mildew, a cluster-cup, a rust, a smut, and a *Synchytrium* as parasites, and we have found three of these infesting the same leaf and the poor thing still alive, if not flourishing!

Certain species of parasitic fungi exist under widely different forms at different seasons, and frequently alternate from one host plant to another—a circumstance which adds greatly to the difficulty of study and identification. Certain forms which were formerly recognized as distinct species, and even as members of different genera are now known to be simply the diverse stages of the same parasite preying upon widely different hosts at different seasons.

Since we have spoken of the evident relation of fungi and algae it is best to make a single reference here to the probable origin of of the fungi.

1. It is clear that the lower types of fungi known as algal-fungi have been derived from their nearest allies among the green plants. Some, doubtless, by continued living where there was

more or less of a supply of organic matter in the water ready to be appropriated, found it easier to absorb than to construct, and gradually lessened their amount of chlorophyl until they came ultimately to live entirely on what they absorbed. As among higher beings the step from independence and productiveness to indolence and beggary was a simple one. In other cases the simple algae doubtless assumed the parasitic condition in order to receive protection from their host; once inside, where an abundant supply of nutrition was present to be had for the taking, the step from parasitic alga with green chlorophyl of its own to a parasitic fungus without chlorophyl was natural and almost inevitable.

2. It is likewise probable that certain of the higher forms of fungi that have lost their sexual methods of reproduction have been derived from the lower fungi. While they have become more differentiated *structurally*, they have lost the only functional activity that would insure them the possibility of rising above mediocrity, and without sexuality and without the power of locomotion, they have left only the possibility of squatter sovereignty and their inferior position in the vegetable world is forever fixed.

3. The near approach of certain of the spore-sac fungi (Ascomycetes) to the red algae renders it highly probable that some of the higher algae have in like manner become physiologically degenerated into some of the higher fungi.

4. Of the conflicting schools of belief respecting the origin of the higher fungi it is likely, as in most cases where there is contention, that both parties are partially right, and that the higher fungi instead of representing a compact group with a common origin, took their rise from several widely different sources.

CHAPTER III

REPRODUCTION, CONSTITUENTS AND HABITS

Structurally a fungus consists of a cell or cells made up of semi-fluid protoplasm surrounded by a thin wall of cellulose.* The cell may be spherical, oval or elongate, or is commonly united with others into a thread-like structure known as a hypha. Many hyphae, more or less entangled, form what is known as mycelium, which may be cobwebby or flocculent or spread out in an irregular compact layer. Most hyphae are formed of many cells by the division of the original cell and the formation of septa or cross partitions. In certain cases no septa are formed and the boundary between the cells is obscure, in which case we have a coenocyte; this condition facilitates the rapid transfer of food material from one place to another. The hyphae may be simple or branched, and as in higher plants the branching may be monopodial or dichotomous. True roots are never produced but frequently root-like structures are developed as hold-fasts to anchor the plant to the substratum on which it grows.

In parasitic forms, irregular tubercular or root-like projections are developed which sometimes penetrate the cell walls of their host, for the purpose of absorbing the nutriment that would otherwise serve as food for the host plant itself; these are called haustoria. Hyphae are sometimes united to form cord-like structures known as mycelial strands and often become matted together into a felt-like mass or are sometimes hardened into a corky, horny or even woody structure to serve as a protection to more delicate exposed parts, particularly in forms that are perennial in their habits.

Reproduction is accomplished in various ways. In unicellular forms it takes place by budding (gemmation) as in the yeast plant, by self-division (fission) as in bacteria, or by free cell for-

---

*This fungus cellulose apparently differs slightly from normal cellulose, yielding less readily to standard tests than the cellulose of green plants.

mation (internal cell division). In one of the great classes of the fungi the spores are produced in membranous sacs called asci (*Pl. 1, f. 17*); in the common mushroom and allied fungi the spores are borne on spicules (*sterigmata*) which rise from large cells known as *basidia* (*Pl. 1, f. 18, 19*). In various groups of fungi reproductive bodies* are produced in a variety of ways, some of the more common being as follows:

1. Sexual forms of reproduction.

   a. By the union of similar elements (conjugation) resulting in the formation of a zygospore (black mould, *Pl. 2, f. 4, 7, 9*).

   b. By the union of dissimilar elements resulting in the formation of an oospore (downy mildew, *Pl. 3, f. 5*).

   c. By the union of dissimilar elements which, followed by the growth of an alternate stage (known as a sporophyte) results in the formation of a sporocarp (powdery mildew, *Pl. 4, f. 5*).

2. Asexual forms of reproduction.

   a. The formation of ciliated swarm spores within the cell by the ordinary process of internal cell division.

   b. The formation of solitary conidia on simple or branching hyphae (downy mildew, *Pl. 3, f. 4*).

   c. The formation of conidia in chains by the successive cutting off of the ends of certain hyphae (green mould, powdery mildew, *Pl. 4, f. 2, 4*).

   d. The formation of sporangia or membranous receptacles containing large numbers of spores (black mould, *Pl. 2, f. 2, 8.*)

   e. The formation of pycnidia or special receptacles of more or less elaborate structure, from the walls of which the conidia are produced (many leaf spot fungi, *Pl. 5, f. 2*).

The end of a hypha which bears conidia is known as a conidiophore; it may be merely of the same thickness as the hypha itself

---

*In Sylloge Fungorum, Saccardo uses different terms for reproductive bodies according to their method of formation and the group in which they occur, *e. g.*, *sporae* in the Hymenomycetes and Gastromycetes; *sporidia* in the Pyrenomycetes and Discomycetes; *sporulae* in the Sphaeropsideae; and *conidia* in the Melanconieae and Hyphomycetes. For the present we shall use only *spores* and *conidia* for these reproductive bodies. It must be remembered that while their function is practically the same, they cannot be regarded as homologous organs since their origin and method of formation is often widely different.

or may be variously enlarged. Spores and conidia are produced singly in rare cases, more commonly in masses, sometimes in prodigious quantities.

The reproductive bodies (spores or conidia) are of various forms. The simplest are one celled and in form may be spherical, (*Pl. 1, f. 1*), oval (*Pl. 1, f. 2, 3, 4*), elongate, allantoid (*Pl. 1, f. 6*), rod-like, worm-like or thread-like (*Pl. 1, f. 14*). Others by a cross-partition become twin or 2-celled (didymoid) (*Pl. 1, f. 7, 8, 9*); others by further parallel cross-partitions become a row of cells (phragmoid) (*Pl. 1, f. 10–13*); others still by a division of cells in more than one plane become many-celled (dictyoid, muriform) (*Pl. 1, f. 15*). Some spores are hyaline or colorless, others are variously colored, usually some shade of yellow or brown.* The spores of agarics range from white, through pink or salmon-colored to rusty yellowish-brown, and on to dark brown and black. *Lepiota Morgani* is an anomalous species with green spores.

Certain spore forms readily characterize special groups of fungi and make their recognition an easy matter ; such are the spores of the ordinary grain rust (*Pl. 6, f. 3*); many spore forms have no special distinctive character. In aquatic forms the spores are frequently provided with cilia or other means of locomotion, and this feature is also present in certain stages of the development of certain parasites of land plants.

CHEMISTRY OF FUNGI. Besides the cellulose that forms the cell walls, the protoplasm of various fungi develops a great variety

---

*In certain groups of fungi, Saccardo, *loc. cit.*, has made use of artificial group-names based on spore characters which should be familiar to any one attempting to use his work :

ALLANTOSPORAE : Spores simple, cylindric, curved.
PHAEOSPORAE : Spores simple, ovoid, brown.
HYALOSPORAE : Spores simple, ovoid, or oblong, hyaline.
HYALODIDYMAE : Spores 1-septate, hyaline.
PHAEODIDYMAE : Spores 1-septate, brownish.
PHAEOPHRAGMIAE : Spores 2–many-septate, brownish.
HYALOPHRAGMIAE : Spores 2–many-septate, hyaline.
DICTYOSPORAE : Spores transversely and longitudinally septate.
SCOLECOSPORAE : Spores rod-like or filiform.
STAUROSPORAE (ASTEROSPORAE) : Spores angular, forked or stellate.
AMEROSPORAE : Spores globose to cylindric, hyaline or colored.

of compounds. These may be briefly summarized under the following groups:

1. Hydrocarbons, including sugar of the glucose type, glycogen, gums (notably lichenin), mannite, and a number of other forms.

2. Organic acids, among which are oxalic, malic, citric and lactic acid.

3. Aromatic acids. Nineteen forms are described by Zopf.*

4. Fats.

5. Ethereal oils.

6. Resins.

7. Coloring matters in great variety. Over thirty forms are described by Zopf.† These include various shades of brown, blues, purples, yellows, reds and greens.

8. Alkaloids, among which muscarin and ergotin are among the best known. The former is found abundant in the fly-agaric, and is probably a common cause of mushroom poisoning. The latter forms the basis of the ergot poisoning in animals, and is one of the deadly drugs of the pharmacists. Zopf‡ enumerates twelve different alkaloids found in fungi.

9. Cholesterin.

10. Albuminoids of various forms.

Besides the ordinary elements of organic food, carbon, hydrogen, oxygen and nitrogen, fungi require sulphur, phosphorus, one of the alkali metals (commonly potassium), and one of the alkali earth metals (more commonly calcium). These food stuffs are obtained (1) From decaying organic substances, or (2) From the living cells of other plants or animals. Fungi which obtain their nourishment by the former method are called saprophytic. Those preying on other organisms are called parasitic. A plant or animals which supports a parasite is commonly called its host.

Fungi require for profuse growth considerable moisture and a moderate degree of warmth. The optimum temperature for the growth of the conidia of the common green mould (*Penicillium crustaceum*) according to Wiesner is $22°$ C., the minimum being $1.5°$-$2°$ C., and the maximum being $40°$-$43°$ C. Other fungi vary a few degrees from these figures. While certain fleshy fungi

---

* Die Pilze, 131-138. Schenk, Handbuch der Botanik, 4 : 401-408.
† *Loc. cit.*, 143-163. (Schenk, *loc. cit.*, 413-433).
‡ *Loc. cit.*, 163-166. (Schenk, *loc. cit.*, 433-436.)

(notably *Collybia velutipes*) will develop in weather that is only a little above the freezing point, it commonly requires warm wet weather to facilitate the growth of an abundant crop of fungi, and in temperate climates the months from July to October are most prolific in the growth of the higher fleshy forms. On the other hand, many of the black fungi (Sphaeriales) develop their ascospores during the winter, the period from February to May being particularly favorable for their collection.

Fungi inhabit nearly every form of matter living or dead. Decaying wood and other vegetable matter, dead flies and fish, saccharine fluids like preserved fruits, greasy bones, food stuffs like bread and cheese, all furnish the medium in which saprophytic species develop and thrive. Parasitic forms live on flies, grasshoppers, fish, birds and even man among the animals; others attack pollen grain, diatoms, pond scums and other algae, and even other fungi, and every spermaphyte from the conifers to the composites has one or more parasitic species living at its expense. We have rusts, mildews, moulds, smuts and leaf spots not only on every known cultivated plant, but on the wild plants of moor, of forest and of bog, their name is legion. We can distinguish three ordinary types of parasitism :

1. That of internal free parasites floating or swimming in the cell sap of plants or the juices of animals. To this group belong many of the bacteria. Allied to these in habit are the unicellular myxos and *Synchytria* which often live inside the walls of a single cell of their host.

2. That of internal fixed parasites forming mycelium within the tissues of other plants and appearing at the surface only for purposes of reproduction. Such are the rusts and smuts of grain and most of the common injurious fungous diseases of cultivated plants. Their presence is shown by discoloration (yellowing or browning) of the tissues of the plant attacked, and often by thickening or other deformation of the tissues, sometimes, even by the formation of galls. In some cases, as in many of the rusts, the affected area is slight and confined to the immediate region about the point where spores are ultimately formed. In other cases, as in some of the smuts, the mycelium of the parasite is more extended throughout the tissues of the plant and the parasite does not disclose its presence in the host until the time when spores are pro-

duced. Of such a type is the ordinary smut of cereals; the spores of the fungus germinate with the germination of the grain and enter the young plant as it emerges from the seed. Little, if any trace of its presence can be discovered until the grain comes to form its spike or panicle, when in place of ovaries the black smutty spores of the parasite are developed.

3. That of external parasites that form mycelium on the surface of leaves or fruit, penetrating the exterior cells for purposes of nutrition. Such are the powdery mildews which form cobwebby or powdery patches on many leaves as those of roses cultivated in greenhouse, lilacs, dandelions and many cultivated plants. Such too are the black forms of *Meliola* found on leaves in tropical and warm temperate regions.

Certain parasitic fungi are peculiar to the host on which they dwell; other parasites seem to thrive equally well on a variety of hosts. In some instances, the species of certain fungus genera affect certain genera or allied genera of the higher plants. The Leguminosae, *e. g.*, are parasitized by rusts belonging to the genera *Uromyces* and *Ravenelia*, tho species of the first genus occur on other plants; the apples and haws (Pomaceae) by the various species of *Roestelia;* the pines and spruces by the species of *Peridermium;* the roses, brambles and Potentilleae by the species of *Phragmidium;* the Ericaceae by species of *Exobasidium;* Juniperus and *Cupressus* by the species of *Gymnosporangium*.

As a tentative arrangement we may recognize among the fungi: (1) A group of plants of simple structure in which the algal characteristics including sexual methods of reproduction are still manifest, and (2) A higher group in which a more complicated structure is commonly associated with a loss of sexual reproduction; as these higher forms manifest two marked methods of spore production, (*a*) Those enclosed in sacs or asci (*Pl. 1, f. 17*), and (*b*) Those developed free on enlarged basidia (*Pl. 1, f. 19*), we may recognize three convenient classes of fungi as follows:

1. PHYCOMYCETES (the lower or algal fungi).
2. ASCOMYCETES (the sac-spore fungi).
3. BASIDIOMYCETES (the basidial spore fungi).

In the first group sexual reproduction is common. In the

second, it occasionally occurs, and in the third, it is not certainly known to exist.

Besides the above classes, there are two groups of low organisms, the bacteria and slime moulds, which are often associated with the true fungi. The tendency at present is to treat them separately. Many botanists, indeed, regard the slime moulds as animals. If treated with the true fungi they would form two classes additional to those above noted and stand below them: the MYXOMYCETES and the SCHIZOMYCETES. It is better, however, to regard the Mycetozoa (slime moulds) as a group of organisms coördinate with the phylum THALLOPHYTA, and the bacteria with their evident close alliance to the blue green algae (*Cyanophyceae*) may be best regarded as forming together a group (SCHIZOPHYTA) coördinate with the true algae on the one hand, and the true fungi on the other.

It must be constantly borne in mind that these larger groups are heterogeneous assortments of plants united together, not so much because of actual phylogenetic relationships as from the possession of certain characters that indicate a real, though somewhat artificial, resemblance.

Since some slight innovations in group names are to be introduced in succeeding chapters, it is, perhaps, well to interpolate a general statement on group nomenclature. *Species* among fungi, as elsewhere, are recognized as the smallest groups of distinct things, such as might have been produced from the same spores or mycelium; associated species with like characters form *genera*. To the genus and species we give the Latin double name as among all other organic forms of life, *e. g.*, *Amanita caesarea*. It is fortunate that for the majority of fungi there are no common or "popular" names, for there is no necessity for a double series. *Amanita caesarea* contains no more letters than "Caesar's mushroom," and is at once a more direct and elegant method of citation. Like genera are united into families which are normally named from some characteristic or representative genus, with the uniform termination—*aceae*, *e. g.*, Agaricaceae from the genus *Agaricus*, Hydnaceae from the genus *Hydnum*. Closely related families are united into orders whose name is likewise derived from the name of a characteristic genus. The ordinal name according to a recent, but important innovation, has the termination—*ales, e. g.*, Mucorales, from the genus *Mucor* of the family Mucoraceae. Orders are the primary subdivisions of classes.*

The germination of fungi may be prevented by the use of certain mineral salts. Those of copper are more commonly used, especially the sulphate, carbonate and acetate, either alone or mixed with other substances. Two pounds of copper sulphate, dissolved in fifty gallons of water, can be used as a spray on vines and trees before the opening of the buds. After the trees or vines are in full leaf the Bordeaux mixture is used. This is made by dissolving six pounds of copper sulphate in a half barrel of water by hanging near the surface in a piece of gunny sack. In another barrel four to six pounds of best stone lime are slaked in a quantity of water, strained, and diluted to a half barrel. The two substances are now turned together slowly, mixed thoroughly with constant stirring and used for a spray. It should be used the same day it is made, always stirring thoroughly just before using. This mixture has been used successfully in a more dilute condition, even up to fifty gallons for the above amounts, or even with only four pounds of copper sulphate and an equal amount of lime. This is used on peaches, plums, pears, quinces and grapes just before the blossoms open and just after the young fruit has set, and among orchard and vine growers is a well-known application for the prevention of many fungous diseases, such as the peach and plum rot (*Monilia fructigena*), pear leaf blight (*Entomosporium maculatum*), and the various mildews, rots and anthracnoses of the grape as well as many others.

Ammoniacal copper carbonate† and a solution of copper acetate‡ are often used as later sprays in the progress of fruit growth ; the latter is specially recommended for spraying peach and plum trees after the fruit has commenced to turn color since it does not dis-

---

* The above system has been announced at Berlin as the standard for German practice, but in the recent publications issuing from the Hof-Museum, there has been glaring inconsistency in their usage. In the following pages an effort is made to secure rigid uniformity.

† Made by dissolving four ounces of copper carbonate in two quarts of ammonia, and adding the solution to fifty gallons of water.

‡ Four ounces of copper acetate in fifty gallons of water.

figure the fruit, while it is equally effective in preventing the growth of the rot.*

In some cases where the seed will endure the heat, seeds are dipped in hot water for a short time before sowing. To be effective in destroying the vitality of the spores the temperature of the water must not fall below 130° F.; if the temperature rises above 135° F. the hot water is liable to injure the seed. This method has been used effectively in preventing the smut in oats and other grains.

In a few cases where the disease is largely confined to the surface, a soaking for a short time in a dilute solution of corrosive sublimate † has been effective in preventing spread of disease. Another fluid for soaking seeds is a solution of potassium sulphide in water.‡ This has also been used effectively for the smut of oats and other grains.

The amount of attention that has been given to the fungous diseases of plants has been very extensive, and in this country has quite revolutionized certain phases of agriculture and horticulture. In 1885 a bureau was established by the Department of Agriculture at Washington for the purpose of studying plant diseases, and now employs several trained botanists for this purpose. Many investigations have also been carried on by the Agricultural Experiment Stations in the various states and the literature good, indifferent, and bad that has accumulated on this subject is enormous.§

---

* Those desiring further information in this direction are referred to Lodeman. The Spraying of Plants. New York, 1896. (Macmillan & Co.)

† Two and one-half ounces dissolved in two gallons of hot water.

‡ One and one-half pounds of potassium sulphide (liver of sulphur) in twenty-five gallons of water.

§ Dr. W. C. Sturgis (Report Conn. Exper. Sta. 1897: 182-222) has published a most valuable index to the diseases of cultivated plants which will serve to acquaint one with the nature and extent of this literature and the effect of treatment looking towards the control of plant diseases.

CHAPTER IV

CLASS I. PHYCOMYCETES

(*The Alga-like Fungi.*)

The PHYCOMYCETES or algal fungi are characterized as follows:

1. The plant body ranges from an undifferentiated mass of protoplasm living parasitically inside a single cell to well developed hyphae with horizontal and vertical branches as seen in the common black mould of bread.

2. The method of reproduction is commonly asexual either by the formation of sporangia or conidia. In some of the lower forms these are the only methods. In the higher forms, sexual reproduction by conjugation or by the normal process of fertilization by antherids is occasionally or in some species commonly found.

3. In habit some are saprophytic, like many of the common moulds,* others are parasitic. A number of well-known diseases of cultivated plants are produced by members of this class of fungi, notably the downy mildew of the grape, the potato rot, and some of the diverse phenomena known as damping off. The fish mould or salmon disease, often very destructive to young fish in hatcheries also belongs to this group of fungi.

The class Phycomycetes contains five well-marked orders arranged in three sub-classes distinguished by their method of sexual reproduction.

1. ARCHIMYCETES. Sexual reproduction rarely developed; mycelium wanting or poorly developed.
    1. **Chytridiales.** (Mostly parasitic on algae.)
2. ZYGOMYCETES. Sexual reproduction by conjugation.
    2. **Mucorales.** (Saprophytic moulds, or parasites on other moulds.)
    3. **Entomophthorales.** (Insect parasites.)

---

* A considerable number of moulds that are more or less common belong to the Moniliales; the common green mould belongs to the Aspergillales.

3. OÖMYCETES. Sexual reproduction accomplished by the fertilization of an egg-cell by an antherid.
  4. **Saprolegniales.** (Mostly aquatic moulds.)
  5. **Peronosporales.** (Parasites on spermaphytes.)

The orders may be more easily distinguished by the following somewhat artificial key, based on more easily noted characters:

Mycelium wanting or poorly developed; sexual reproduction usually wanting; parasites on algae, protozoans or rarely on spermaphytes.
                CHYTRIDIALES.
Mycelium well developed.
 Asexual reproduction by zoöspores; aquatic moulds (one species parasitic on seedlings).   SAPROLEGNIALES.
 Asexual reproduction by aerial conidia or sporangia.
  Parasitic on spermaphytes.   PERONOSPORALES.
  Parasitic on insects.   ENTOMOPHTHORALES.
  Saprophytic moulds or parasitic on other moulds. MUCORALES.

Order 1. CHYTRIDIALES.

The Chytridiales are parasitic plants of low organization often confined to a single cell. They prey for the greater part on infusoria and other protozoans, desmids, diatoms (*cf. Pl. 2, f. 1*), filamentous algae and other fungi. A few live in pollen grains, and others, notably members of the genus *Synchytrium*, are parasitic in the foliage of the higher plants.

The species of *Synchytrium* have no mycelium, the plant body consisting of a single cell living parasitic either in a single epidermal cell of its host which it greatly enlarges into a gall (*Pl. 2, f. 2*), or forming a gall from a number of cells (*Pl. 2, f. 1*). The parasite either becomes a resting spore or a sporangium. In either case reproduction is accomplished by the formation of zoöspores. A common species is *S. decipiens* which is found on the leaves of *Falcata* during the entire season, and bears a superficial resemblance to a red rust (*Uredo*) in which genus in fact, it was originally described. An allied species is found on *Oenothera*. One of the species parasitic in a single cell is found on *Anemone* in early spring, forming little reddish papilliform galls on the upper surfaces of the leaf.

Certain of the members of this order do not form a cell wall even, while some develop a sort of imperfect mycelium. In none of them is there anything more than a rudimentary form of sexuality.

There is much difference of opinion in the matter of classification. Over two hundred species arranged in some forty genera are usually recognized. Various members of this group show affinities with the *Myxomycetes* and others with certain forms of Basidiomycetes (*Ustilaginales*). Except *Synchytrium*, the American forms of this order have scarcely been studied.

LITERATURE.

Fischer. Rabenhorst's Kryptogamen Flora Deutschland, Oesterreich und der Schweiz. 1^4: 3–5, 11–160. 1892, is the best monograph of these forms. A somewhat different generic arrangement appears in

Schroeter. Die natürlichen Pflanzenfamilien, 1^1: 64–92. 1892–3. In both of these works citations of the extended but widely scattered European literature will be found.

Descriptions of the species will also be found in

Saccardo. Sylloge Fungorum, 7: 286–322; 9: 357–363; 11: 246–251.

The American forms of *Synchytrium* are described in

Farlow. The Synchytria of the United States. Bot. Gaz. 10: 235–245. *Pl. 4*. 1885.

Order 2. MUCORALES.

The order Mucorales includes the more common moulds which are found on various organic substances, with the exception of the ordinary green mould which belongs to another order of fungi. These true moulds are either saprophytic on various organic substances or are parasites on other moulds. The common *Mucor stolonifer** will illustrate well the habit and structure of the order. The plant body consists of irregular branching hyphae from which arise lateral branches which form at intervals two sorts of secondary branches, the one descending and forming root-like holdfasts, the others erect and bearing sporangia. These sporangia contain a large number of conidia which are dark colored at maturity and give the ripe sporangium a black appearance. These conidia asexually produced are the ordinary method of reproduction in

---

*Also known under the name of *Rhizopus nigricans*.

the plant, and they retain their germinating power for a long time. Sexual reproduction takes place by the advance of two short hyphal branches toward each other which unite and fuse after each has formed a special reproductive cell. As a result of this union, a zygospore with a thick protective covering is borne on the united hyphae which are then known as suspensors. (*Pl. 2, f. 2, 3, 4.*)

Another common mould growing mostly on horse dung (*Pl. 2, f. 8, 9*) illustrates a different type of growth. The creeping mycelium sends up slender crystalline columns each of which forms a single urn-shaped body at the summit ending in a globular or lens-shaped black sporangium. By the accumulation and ultimate compression of the gases of decomposition in this urn, the sporangium is shot upwards with considerable force and with a slight report. By holding the hand palm downward over a mass of these mature moulds, the discharge can be distinctly felt and the sporangia are sometimes propelled to the distance of several feet. This fungus is known as *Pilobolus crystallinus*.

Other genera can be found growing on moulds which have been for some time under cultivation, *Chaetocladium Jonesii* is one of the common species and *Piptocephalis Friesii* (*Pl. 2, f. 5, 6, 7*) is another frequently found.

In some genera the sporangium becomes reduced in size and its contents instead of dividing to form numerous spores, remains entire and so have the appearance of simple conidium; these conidia-like sporangia are sometimes borne singly, sometimes in clusters. In some genera the suspensors produce branches after conjugation which form a sort of protective covering to the zygospore. In most cases the sexual reproduction occurs on creeping hyphae; in *Sporodinia*, a form comparatively common in summer on the larger fleshy fungi (notably the *Boleti*), the conjugation occurs on aerial hyphae.

The moulds can be easily developed on various kinds of media, like dung of various sorts (fresh horse dung is the simplest and most thoroughly productive), and form an exceedingly interesting group for study. Much has yet to be learned of the development of many of the species, and while many forms are known to occur in America, only a few of our forms have been studied with any degree of care.

The families and genera of the Mucorales can be distinguished by the following table, adapted and abridged from Schroeter:

1. Asexually-formed spores exclusively or predominately developed in sporangia which sometimes resemble conidia............................ 2.
   Asexually-formed spores produced as conidia; sporangia only exceptionally developed. (Species parasitic on other moulds.)..........11.

2. Sporangia (at least the principal ones) with columella; conidia formation wanting or slightly developed; zygospore naked or covered with a loose growth from the suspensors. **Fam. 1. Mucoraceae.** 3.
   Sporangia without columella; conidia present; zygospores encased in a thick covering. (Species parasitic on other moulds).
   Fam. 3. Mortierellaceae. 10.

Family 1. Mucoraceae.

3. Sporangium membrane uniformly developed, not cuticularized, breaking or melting away. 4.
   Sporangium membrane cuticuliarized in the upper half and persistent, thin and melting away in the lower half. 9.

4. Sporangia all similar furnished with columella (only exceptionally in small lateral sporangia). 5.
   Sporangia of two kinds, principal and secondary; principal at the end of hyphae always with a columella. 8.

5. Sporangiophores simple or branched but not repeatedly dichotomous; zygospores formed on the horizontal mycellium. 6.
   Sporangiophores repeatedly dichotomous; zygospores on aerial hyphae. SPORODINIA.

6. Suspensors with no outgrowth at maturity. MUCOR.
   Suspensors with a thorny outgrowth at maturity. 7.

7. Thorns of suspensor projecting; zygospore naked. PHYCOMYCES.
   Thorns closed over the zygospore forming a loose covering. ABSIDIA.

8. Principal sporangia with columella, secondary without; spores uniform in both THAMNIDIUM.
   Principal sporangia with columella, those of the secondary tongue-shaped; spores different in two kinds of sporangia. DICRANOPHORA.

9. Sporangiophores uniformly cylindric under the sporangium; sporangia not discharged. PILAIRA.
   Sporangiophores swollen under the sporangium; sporangia forcibly discharged at maturity. PILOBOLUS.

Family 2. Mortierellaceae.

10. Sporangiophores erect ; branches tapering at the apex. MORTIERELLA.
    Sporangiophores twining, forming numerous lateral branches everywhere equally thick.   HERPOCLADIELLA.
11. Conidia formed singly (*i.e.*, not in chains) ; zygospores formed directly from the gametes. **Fam. 3. Choanophoraceae.** 12.
    Conidia in chains ; zygospores developed at the apex of the arched gametes.   **Fam. 5. Piptocephalidaceae.** 13.

Family 3. Choanophoraceae.

12. Sporangia present (One genus of East India species). CHOANOPHORA.

Family 4. Chaetocladiaceae.

Sporangia wanting (a somewhat common genus). CHAETOCLADIUM.

Family 5. Piptocephalidaceae.

13. Ends of the conidiophores on which the basidial cells rest of the same thickness as the branches.   PIPTOCEPHALIS.
    End of the conidiophores enlarged, capitate.   14.
14. Conidiophores not branched.   SYNCEPHALIS.
    Conidiophores branched.   SYNCEPHALASTRUM.

Of the above, *Mucor* is the largest genus with some fifty species. With the exception of *Thamnidium*, ten species, *Pilobolus*, seven species, *Mortierella*, sixteen species, *Piptocephalis*, eight species, and *Syncephalis*, seventeen species, most of the genera are small, several of them consisting of a single species.

LITERATURE.

The literature of the most value in the systematic study of this group is :

Fischer. *Loc. cit.* 6, 7, 161–310.

Schroeter. *Loc. cit.* 119–134.

Brefeld. Botanische Untersuchungen über Schimmelpilze, 1 : 1872 ; : 1881.

Saccardo. Sylloge Fungorum, 7 : 182–233 ; 9 : 335–340 ; 11 : 239–242.

Van Tieghem et Le Monnier. Recherches sur les Mucorinées. Ann. Sc. Nat. V. 17 : 261–399. *Pl. 20–25.* 1873.

Van Tieghem. Nouvelles recherches sur les Mucorinées. Ann. Sc. Nat. VI. 1 : 1–175. *Pl. 1–4.* 1875.

Van Tieghem. Troisième mémoire sur les Mucorinées. Ann. Sc. Nat. VI. 4 : 312-398. *Pl. 10-13.* 1876.

Some of the American species have been catalogued by

Pound. A Revision of the Mucoraceae with especial Reference to Species reported from North America. Minn. Bot. Studies, 1 : 81-104. 1894.

Order 3. ENTOMOPHTHORALES.

The Entomophthorales are parasitic on flies, grasshoppers, the larvae of beetles and other insects. They sometimes produce destruction of insects in large numbers. The common *Empusa muscae* or fly fungus is one of the most familiar examples of this order. It is a common thing in autumn to see flies with white swollen bodies hanging dead on walls and windows with a white radiant halo surrounding them. These are the victims of the fly fungus. Within the body is the mycelium of the fungus producing the death of the victim and developing simple conidia which are thrown to some distance and form the halo. Sexual reproduction takes place by conjugation and in some species only zygospores are produced. In late summer it is common to see grasshoppers wearily crawl up the stems of mulleins or thistles, turn black in the face and die, clinging with a death grip to the stem on which they have climbed. These insects are full of the mycelium of a mould which has been slowly sapping their vitality. After their death, spores are formed which are more easily distributed because of the last act of the insect seeking an elevated position.

A third instance is seen in a parasite of the clover weevil. The fungus affects the insect in the larval condition, in which state the insect feeds on the roots of clover ; when a certain stage in the growth of the fungus is reached, the larva becomes weary of life and lazily crawls up a blade of grass, coils horizontally around the tip (*Pl. 3. f. 7*) and dies ; from his elevated position spore dissemination is simple. Over fifty species are known belonging to seven genera; four-fifths of the species belong to the genera *Empusa* and *Entomophthora*. The American forms have been carefully studied by Thaxter.

LITERATURE.

Schroeter. *Loc. cit.* 134-141.
Saccardo. Sylloge Fungorum, 7 : 280-286 ; 9 : 349-357.

Brefeld. Botanische Untersuchungen über Schimmelpilze, **4** : 1881; **6** : 1884.

Thaxter. The Entomophthoreae of the United States. Mem. Boston Soc. Nat. Hist. **4** : 133–201. *Pl. 14–21.* 1888.

Order 4. SAPROLEGNIALES.

The order Saprolegniales contains mostly aquatic moulds which live on dead flies, dead fish, or quite commonly attack living aquatic animals, commonly young fish. They have been called fish-moulds, and the disease they engender has been known in England as the salmon disease. Sometimes they become very conspicuous, forming branching masses two or three inches. high. They are commonly much less conspicuous, however. Certain species are capable of living in the tissues of various plants, one of which, *Artotrogus DeBaryanus*,* is a common cause of the phenomenon of damping off, commonly known to gardeners as a destructive disease of young seedlings, though several other fungi will produce a similar disease. In the truly aquatic species the asexual reproduction is accomplished by zoöspores. In the semi-aerial forms (Pythiaceae),† the asexual reproduction as accomplished by conidia. The sexual method of reproduction in both families is accomplished by means of the fertilization of egg-cells by antherids.

LITERATURE.

Schroeter. *Loc. cit* 93–197.

Fischer. *Loc. cit.* 310–383.

Saccardo. Sylloge Fungorum, **7** : 264–280 ; **9** : 345–349 ; **11** : 244–245.

Humphrey. The Saprolegniaceae of the United States, with notes on the other species. Trans. Amer. Philos. Soc. **17** : 63–148. *Pl. 14–20.* 1892.

Thaxter. Observations on the genus *Naegelia* of Reinsch. Bot. Gaz. **19** : 49–55. *Pl. 5.* 1894.

Thaxter. New or peculiar aquatic fungi. Bot. Gaz. **20** : 433–440, *Pl. 29;* 477–485. *Pl. 31.* 1895.

---

* Formerly known as *Pythium DeBaryanum.*

† Fischer (*loc. cit.*), with apparently good reason unites the conidia-bearing Pythiaceae with the Peronosporales and also unites the aquatic Monoblepharidaceae with the Saprolegniales.

Atkinson. Damping off. Bull. Cornell Univ. Agric. Exper. Sta. 94 : 233-272. *Pl. 1-6.* 1895.

Order 5. PERONOSPORALES.

The Peronosporales are parasitic plants preying mostly on the higher plants and form an extensive and destructive group of fungi. There are two types representing two families. One type commonly known as white rust, forms milk-white glistening clusters of conidia under the epidermis of shepherd's purse and other crucifers, and the other forms whitish mould-like masses of branching conidiophores on the under surface of leaves of various plants. The latter type includes two of the most destructive diseases of cultivated plants affecting the grape and the Irish potato respectively, and commonly known as downy mildews. The species infesting the potato has been known to destroy from one-third to one-half of the entire crop of potatoes of a whole state ; and the one affecting the grape after its introduction into France in 1878 caused consternation among the vineyard owners of southern Europe until methods of treatment were found for the prevention of its ravages. In this country it caused the abandonment of numerous vineyards during the same period.

The downy mildew of the grape attacks all green portions of the vine, its mycelium growing as an internal parasite and drawing its nourishment by means of haustoria (*Pl. 3, f. 3*). The symptoms are :

1. Yellowish spots on the upper surface of the leaf with mould-like conidiophores on the corresponding lower surface, the spots ultimately becoming brown and dead. In bad cases the entire leaf may become involved in which case it soon shrivels as if burned.

2. The berries if attacked early do not attain one-fourth their normal size and turn brown or gray if conidia are formed. If attacked later the full-grown berries will show the same colors. This is sometimes known as the brown rot, but must be clearly distinguished from the black rot, which is due to an entirely different parasite.

The downy mildew of the grape, in common with other species of the order, produces two kinds of spores which are different in origin, form and function.

1. *Conidia.* These are asexually produced on branching conidiophores which project from the stomata on the under side of the leaf (*Pl. 3, f. 4*). They germinate soon after they are produced and their function is to rapidly spread the fungus during favoring conditions. For this reason they are often called summer spores. In germinating, the contents of the conidium break up into a number of zoöspores which swim in the moisture covering the leaf like a film, sprout, penetrate the leaf through one of the stomata and develop new centres of growth. The distribution of the conidia is aided by the wind.

2. *Oöspores.* These are produced within the host by the fertilization of an egg-cell of an oögone by a smaller organ known as an antherid (*Pl. 3, f. 5*). This results in the formation of a new cell, which becoming surrounded with a thick wall and is known as the oöspore. From the fact that this remains dormant some time before germination it is often called a resting-spore, since its function is to carry the life of the fungus over an unfavorable period, it is also called a winter spore. It remains within the tissues of its host until the spring or until their decay occurs when it germinates after a manner similar to the conidia.

The remedy for the disease is found in spraying the vines with copper salts to prevent the germination of the summer spores.

The rot of the Irish potato is due to a similar fungus which produces a wilt of the leaves, accompanied by the formation of conidia similar to those of the grape. These conidia, however, are produced in a slightly different manner, the conidiophore producing branches only after the terminal conidia are developed. The disease extends to the tubers, forming brown or black discolorations, which end in the extensive decay of the entire tuber. The mycelium is perennial, living from year to year in the tuber, and will break out into an epidemic whenever favorable conditions occur.

The disease is to be controlled more by the judicious selection of seed and land and the proper destruction of diseased potatoes, though spraying may prevent the spread of the disease by means of the conidia.

The families and genera of the Peronosporales may be distinguished by the following table :

Family 1. Albuginaceae.

1. Conidiophores club-shaped, formed under the epidermis of the host; conidia formed in chains in white masses under the epidermis of the host. ALBUGO*.
Conidiophores formed outside the epidermis of the host; conidia formed singly, never in chains. 2.

Family 2. Peronosporaceae.

2. Conidia forming zoöspores, or at least discharging their contents as a whole. 3.
Conidia germinating with a primordial membrane. 6.

3. Conidiophores simple up to the formation of the first conidia, later producing lateral branches and conidia. PHYTOPHTHORA.
Conidiophores with conidia-bearing branches, completely formed before the development of the oöspores. 4.

4. Conidiophore formed of a single hypha, which bears small uniform branches on the swollen end. BASIDIOPHORA.
Conidiophores with branches rising from different parts. 5.

5. Oöspores grown fast to the walls of the oögone. SCLEROSPORA.
Oöspores lying free in the oögone. PLASMOPARA.

6. Conidia with a papilla on the upper end through which the primitive hypha develops. BREMIA.
Conidia without papilla, germinating from the side. PERONOSPORA.

Of the above genera *Albugo* has six species. Besides *A. candida** growing on various Cruciferae, we have *A. portulaccae* on purslane; *A. ipomoeae-panduranae* on sweet potatoes and various species of Convolvulaceae; *A. tragopogonis* on thistles, salsify and a few other Compositae; *A. bliti* on *Amarantus* widely distributed, and *A. platensis* on various Nyctaginaceae, more or less common in New Mexico.

Phytophthora has *P. infestans* already discussed on the potato, and other Solanaceae, and *P. phaseoli* on Lima beans. *Basidiophora* has a single species on *Erigeron*, *Aster* and *Solidago*. *Sclerospora* has a single species on various grasses. *Plasmopara* has *P. viticola* already discussed on various forms of native and cultivated grapes, *P. geranii* on the wild geranium (*G. Carolinianum*),

---

* Until recently more commonly known under the later name of *Cystopus candidus*.

P. pygmaea on various Ranunculaceae, *P. Halstedii* on various Compositae, *P. australis* on Cucurbitaceae, *P. obducens* on the cotyledons of seedling *Impatiens*, and several other species on various hosts. Of the forms which germinate without zoöspores, *Bremia* has a single species parasitic on lettuce and allied plants, while *Peronospora* has numerous species parasitic on a wide range of host plants.

LITERATURE.

Fischer. *Loc. cit.* 383-489.
Schroeter. *Loc. cit.* 108-119.
Saccardo. Sylloge Fungorum, 7 : 233-264; 9 : 340-345; 11: 242-244.

The following include more or less complete enumerations, descriptions and notes on American species :

Farlow. Enumeration of the Peronosporeae of the United States. Bot. Gaz. 8 : 305-315 ; 327-337. 1883.

Farlow. Additions to the Peronosporeae of the United States. Bot. Gaz. 9 : 37-40. 1884.

Swingle. Some Peronosporaceae in the Herbarium of the Division of Vegetable Pathology. Jour. Mycol. 7 : 109-130. 1892.

The Phycomycetes may properly be regarded as the descendants of degenerate algae which have lost their power of independent existence through parasitism. They have probably descended not from any single ancestral algal stock, but different members have risen from various parent stocks.

CHAPTER V

CLASS II. ASCOMYCETES

(*The Spore-sac Fungi.*)

The members of the second class of Fungi, the ASCOMYCETES, are characterized by their method of producing spores in delicate membranous sacs called asci. These asci, at least in the higher forms are collected together in a body which may be spherical, flask-like, cup-shaped or disc-like; this body is variously known in the different orders as a perithecium, an ascoma, a receptacle or an apothecium. We may speak of it in general as an ascocarp. Mixed with the asci are simple or branched bodies known as the paraphyses. (*Pl. 4, f. 10.*) The ascomycetes vary greatly in size, habit and structure; mycelium is usually abundantly developed and is usually branched and with cells clearly marked by transverse septa or partition walls. The mycelium is arachnoid (cobwebby) in a few forms, but in most species it is usually concealed or buried, while the spore-producing body becomes the conspicuous feature of the plant. In size the members of this class of fungi range from one-celled organisms, like the yeast plant floating in a liquid medium, to large fleshy forms several inches in height, or in some cases hard, almost woody structures are developed, forming conspicuous masses. Most, however, are inconspicuous members of the plant world. Many of them are saprophytic, living in rich soil, or more commonly on dead stems or trunks of trees; a few are subterranean; a number of fleshy forms are edible, like the morel and truffles. A large number of forms are parasitic and produce a variety of diseases among cultivated plants like the leaf curl of the peach, plum pockets, the black knot of the cherry and plum, and the various powdery mildews.

Besides the ascospores, many species produce conidia which are developed in a great variety of ways; some also produce pycnidia and occasionally other forms of reproductive bodies are developed. In this way a single species may have four or five forms of spores, with different shapes, functions, methods of production and periods of germination. In some instances the conidia are produced on one host and the ascocarps on another, the former being frequently parasitic and the latter saprophytic. A large number of conidia- and pycnidia-bearing species is known whose relation to ascus-bearing species is unknown ; in some cases it is probable that the ascosporic condition does not exist. These forms have commonly been called *Fungi imperfecti* from their real or apparent resemblance to the conidial and pycnidial forms of ascus-bearing fungi. It is more than probable that many forms that have been classified in this convenient catch-all are species perfect in themselves.

Counting the *Fungi imperfecti* there are seventeen orders of Ascomycetes all but one of which are well represented in America.* These orders may be distinguished by means of the following synoptical table :

1. Asci separate from each other, not uniting in a special perithecium or ascoma, and without a special covering. 2.

    Asci grouped or fasciculate, surrounded by a spherical, cylindric or pyriform shell or perithecium. 3.

    Asci collected in a hymenial layer, remaining enclosed in a tuber-like ascoma ; habit subterranean. (Truffles.)   10. **Tuberales**.

    Asci collected in a flattened or concave hymenial layer (ascoma) often bordered by a distinct layer. (DISCOMYCETES.) 8.

2. Asci entirely isolated or formed at different parts of the mycelium which is often undeveloped ; vegetative reproduction accomplished by gemmation. (Yeast plants.)   2. **Saccharomycetales**.

    Asci approximate and forming an indefinite hymenium. (Our forms are mostly parasitic, deforming their hosts, forming bladder plums, leaf curls, etc.)   3. **Exoascales**.

3. Perithecia borne on a short pedicel ; microscopic fungi parasitic on beetles and other insects.   9. **Laboulbeniales**.

    Perithecia sessile, either solitary and free, or united and imbedded in a stroma. 4.

---

*The Protomycetales (*Hemiasci*) are low forms of fungi with usually a many-spored ascus. Of these, one form, *Protomyces polysporus*, has been reported from this country parasitic on *Ambrosia trifida*. For literature on the three families of this order see Schroeter, *loc. cit.* 143-149.

4. Asci arranged at different levels in the perithecium, sometimes forming skein-like masses.   **4. Aspergillales.**
Asci in fascicles arising from a common level. (PYRENOMYCETES.) 5.

5. Perithecia globose, scattered, without apparent ostioles, mostly attached to an apparent mycelium or membrane; or flattened and ostiolate in one family.   **5. Perisporiales.**
Perithecia with distinct ostioles.   6.

6. Perithecia (and stroma if present) fleshy or membranous, bright colored (white, yellow, red, or blue).   **6. Hypocreales.**
Perithecia (and stroma if present) hardened, never fleshy, rarely membranous, dark-colored (black or dark brown).   7.

7. Walls of perithecia scarcely distinguishable from the stroma.
    7. Dothideales.
Perithecia with distinct walls, either free or imbedded in a stroma.
    8. Sphaeriales.

8. Ascoma more or less completely closed at first, opening free at maturity, and plane, concave, or rarely convex. (Cup fungi.)   9.
Ascoma open from the first, normally clavate or convex, or pitted, or with gyrose furrows. (Morels, lorchels.)   **14. Helvellales.**

9. Ascoma long enclosed in a tough covering which becomes torn open at the maturity of the spores.   10.
Ascoma soon becoming free, without special covering; mostly fleshy cup-like fungi.   **13. Pezizales.**

10. Ascoma mostly elongate, the cover opening by a longitudinal fissure.   **11. Hysteriales.**
Ascoma roundish, the cover rupturing by radiating or stellate fissures.
    12. Phacidiales.

Order 2. SACCHAROMYCETALES.

This group of organisms includes the yeast plants together with a few other low fungi which up to this time have not been reported from this country. The yeast plant is concerned in the process of alcoholic fermentation and is equally useful in the manufacture of bread and beer. Yeast plants are unicellular, grow in saccharine solutions and reproduce by gemmation (*Pl. 4, f. 1*). Under rare circumstances, mostly when there is a sudden diminution of their food supply,* they reproduce endogenously, a

---

* This condition can readily be secured by removing "top yeast" from a solution in which growth is rapid, and placing it on a moist slab of plaster paris or a fragment of unglazed porous earthenware. The endogenous spore formation ought to be seen in forty-eight hours or less.

condition which by some is supposed to ally them to the ascomycetes of which they are commonly regarded a low degraded form; there is however little more reason for considering these as asci than as sporangia. The spores thus produced possess greater powers of resistance to desiccation; they germinate to form budding yeast cells. The yeast plant is the active agent of all forms of yeast, tho various species of bacteria are usually mixed with them and share in the function of raising bread. The relation of the various forms of yeast to the process of brewing has become a subject of special study. Besides the family Saccharomycetaceae, containing the yeast fungi, the order contains the family Endomycetaceae with four genera which have heretofore been reported only from Europe.

LITERATURE.*

Schroeter. Die natürlichen Pflanzenfamilien 1^1: 150–156.
Saccardo. Sylloge Fungorum, 8: 916–922.
Zopf. Die Pilze, 411–425. 1890.
Bay. The spore-forming Species of the Genus Saccharomyces. Amer. Nat. 27: 685–696. 1893.
Golden. Fermentation in Bread. Bot. Gaz. 15: 204–209. 1890.
Pasteur. Studies on Fermentation. (English translation.) Macmillan & Co. 1879.

Order 3. EXOASCALES.

The order Exoascales includes a number of parasitic fungi that attack various plants, notably the drupaceous fruit trees. One of the most common produces the "peach curl" frequently seen in the young leaves of the peach. In this disease the leaves become variously distorted and deformed and are covered with patches of a reddish or sometimes a whitish color. In badly affected leaves the leaf finally withers and shrivels up as if burned. Other species are found on various plum and cherry trees producing deformities

---

*The technical literature bearing on yeast as an economic factor in the production of beer is enormous, and the continuous study of its conditions and physiological activities constitutes the work of specialists in every large brewery.

in the fruit known as bladder plums, and occasionally on the young branches.

The plants of this order are characterized by the production of asci separate from each other, usually standing side by side on the surface of the affected leaf or other portion of the host.

The families and genera may be distinguished by the following :

Family 1. Ascocorticiaceae.

1. Saprophytic ; asci standing close together on a basal membrane. A single genus and species. ASCOCORTICIUM.

Family 2. Exoascaceae.

Parasitic ; asci free from each other, breaking out directly from the surface of the host with no distinct membrane beneath. 2.

2. Asci formed as swellings at the ends of mycelial threads which project between the cells of the host. MAGNUSIELLA.
Asci springing from a more or less developed subcuticular mycelial layer. 3.

3. Asci eight- (or sometimes four-) spored ; mycelium perennial.
EXOASCUS
Asci many-spored through the budding of the spores within the ascus.
TAPHRIA.

Magnusiella contains two American species producing deformities, the one on *Potentilla Canadensis* and the other on *Betula populifera*. *Exoascus* has numerous species and includes those producing deformities in drupaceous fruits. Most common are *E. deformans* causing the leaf curl, and the various species producing "bladder-plums" or "plum-pockets" from the ovaries of various species of plum and cherry. *Taphria** contains several species parasitic on *Quercus*, *Populus*, *Rhus copallina*, etc.

LITERATURE.

Schroeter. *Loc. cit.* 156–161.
Saccardo. Sylloge Fungorum, 8 : 811–820 ; 10 : 67–72 ; 11 : 435–439.

---

*This is the original form of the generic name. It was changed to *Taphrina* because of the existence of a genus *Taphria* in zoology. This limitation in nomenclature no longer exists, hence the return to the original name.

Sadebeck. Die parasitischen Exoasceen. Jahrb. wiss. Anstalten Hamb. 10 : 1–110. *Pl. 1–3.* 1893.

Schroeter. Die natürlichen Pflanzenfamilien, 1¹ : 156–161.

Robinson. Notes on the genus *Taphrina*. Ann. Bot. 1 : 163–176. 1887.

Atkinson. Notes on some Exoasceae of the United States. Bull. Torr. Bot. Club, 21 : 372–379. 1894.

———— Leaf curl and Plum pockets. Bull. Cornell Univ. Agric. Exp. Sta. 73 : 319–355. *Pl. 1–20.* 1894.

Patterson. A study of North American parasitic Exoasceae. Bull. Lab. Nat. Hist. Iowa State Univ. 3 : 89–135. *Pl. 1–4.* 1895.

Order 4. ASPERGILLALES.

The fungi associated in this order are plants of very diverse habits. One group is made up of subterranean fungi which resemble the truffles. Among the commonest of these are species of *Elaphomyces* about the size of hickory nuts which grow two or three inches below the surface of the soil ; these are often parasitized by a species of *Cordyceps* which projects above the ground and reveals the presence of the *Elaphomyces* which otherwise makes no visible sign.

The second group contains the common green mould (*Penicillium crustaceum*) which grows on decaying or preserved fruit, cheese or other forms of organic matter. The green, dusty portion represents the conidial stage (*Pl. 4. f. 2*) while the ascosporic stage is found in small ascocarps which develop as the probable result of sexual reproduction (*Pl. 4. f. 3*).* With this family *Meliola* is included, a genus largely represented in the South and in tropical regions generally, which has an external mycelium similar to the Erysibaceae but black, often forming large areas on leaves. The relations of *Meliola* and some other allied genera is, however, doubtful. The order is known as the Plectascineae of the Engler-Prantl revision. We have six families as follows :

1. Peridium made up of loose floccose hyphae. **Gymnoascaceae.**
   Peridium compact, closed.   2.

---

*Similar to this is the common herbarium mould (*Aspergillus herbariorum*) which frequently develops on plants under pressure for the herbarium when the driers are not frequently changed.

2. Ascocarps subterranean, mostly enlarged, tuberous. 3.
Ascocarps not subterranean, mostly small. 4.
3. Peridium clearly distinct from the walls of the ascocarp; spore masses
powdery at maturity. **Elaphomycetaceae.**
Peridium not clearly limited, continuous with the walls of the ascocarp;
spore masses never powdery. **Terfeziaceae.**
4. Ascocarps mostly sessile; peridia usually remaining closed.
Aspergillaceae.
Ascocarps mostly stalked; peridia opening at maturity by lobes or
irregularly. **Onygenaceae.**
Ascocarps sessile, the spore masses exuding in columnar masses from
the goblet-shaped peridia. **Trichocomaceae.**

Of the above families, the Elaphomycetaceae, Onygenaceae and Trichocomaceae are each made up of a single genus from which the respective families receive their names. The subterranean Terfeziaceae are made up of eight genera, while the mould-like Gymnoascaceae and Aspergillaceae are more numerous, the former with five and the latter with fourteen genera.

LITERATURE.

Saccardo. Sylloge Fungorum, 1: 60–71; 8: 863–872; 9: 413–431; 11: 441.

Fischer. Die natürlichen Pflanzenfamilien, 1¹: 290–320.

Ellis & Everhart. North American Pyrenomycetes, 45–49. 1892.

Martin. Synopsis of the North American Species of Asterina, Dimerosporium and Meliola. Jour. Myc. 1: 133–139, 145–148. 1885.

Order 5. PERISPORIALES.

The Perisporiales form a small group of partly parasitic and partly saprophytic fungi some of which form a simple and easily studied group, as they are abundant and widely distributed. To illustrate the general characters of this group we can do no better than to describe the structure of the powdery mildews. These are the most typical examples of external parasites and take their popular name from the white powdery conidia that are produced early in the season. The parasite forms masses of cobwebby mycelium on the surfaces of various leaves. A common example is seen on the leaves of the common lilac which is almost universally covered with one of the species of this order. Some species sometimes appear on young stems and fruit as well as leaves. One species is common on the peculiar knot-like fascicles of twigs that disfigure the hackberry tree of the Mississippi valley region, and another is confined to the brown rust-like masses common on the leaves of the beech. The fungus draws nourishment from its host by means of haustoria, which are mere expansions of its hyphae. Two kinds of reproductive bodies are present : (1) Conidia, which are produced by the successive cutting off of the ends of erect hyphae. (*Pl. 4. f. 4.*) These constitute the summer spores, and are blown to other leaves, germinate quickly and rapidly spread the growth of the parasite during the growing season. (2) Perithecia, which are spherical bodies with a thickened protective wall and contain the ascospores enclosed in membranous sacs (asci). (*Pl. 4. f. 5.*) These first appear as minute whitish bodies soon changing to yellow, brown and finally black ; they can be easily seen with a hand lens ; all stages can frequently be seen at the same time on one leaf. It has been supposed that these perithecia are produced as the result of sexuality. (*Pl. 4. f. 6, 7,* 8.) In any event the ascocarp is developed after a plan similar to the homologous organ in some of the higher algae where sexuality is the exciting cause. The ascospores remain over winter and germinate the following spring.

The perithecia are mostly provided with a series of appendages which have different forms and serve in part as the basis of separation of genera. Some of these appendages are needle-like and provided with a bulbous enlargement at their base, some are coiled or hooked at the end, others are dichotomously divided often into an elaborate pattern, others still are similar to the ordinary mycelium, tho often differently colored.

One species produces the common powdery mildew of the grape, which is also very common on the Virginia creeper ; another produces a common disease on young cherry and plum trees ; another forms the destructive hop-mildew, a fourth attacks wheat and other grasses, and still another the gooseberry. There are species found on the maple, elm, oak, basswood, hackberry and numerous other trees ; on the lilac, willow, huckleberry, haw and various other shrubs, and on a great variety of herbaceous plants commencing with the dandelion and the cockle-bur, and including hosts from widely separated orders. They can usually be readily recognized by the cobwebby mycelium on the surface of the leaf and are not likely to be mistaken for any other fungus; in some species, however, the mycelium is not conspicuous and so some common species are often passed by; this is notably true of the species growing on basswood. Some species live on a wide range of hosts, while others are seemingly confined to a single host plant.

The Perisporiales are represented by three families which can be distinguished as follows:

1. Perithecia mostly spherical, imperforate.   2.
Perithecia flattened, shield-shaped, ostiolate.   **Microthyriaceae.**
2. Mycelium external, white; perithecia with appendages. **Erysibaceae.**
External mycelium dark-colored or wanting; perithecia without appendages.   **Perisporiaceae.**

The genera of the Erysibaceae can be readily distinguished by the following synopsis :*

1. Spores one-celled, hyaline.   2.
Spores muriform; appendages similar to the mycelium or wanting.   SACCARDIA.
2. Appendages needle like, enlarged at base.   PHYLLACTINIA.
Appendages dichotomous at the apex.   3.
Appendages hooked or coiled at the apex.   UNCINULA.
Appendages indeterminate, similar to the mycelium.   4.
3. Perithecia containing a single ascus.   PODOSPHAERA.
Perithecia containing several asci.   MICROSPHAERA.
4. Perithecia containing a single ascus.   SPHAEROTHECA.
Perithecia containing several asci.   ERYSIBE.†

The genus *Saccardia* is represented with us by a single species from Florida on *Quercus laurifolia;* the other genera contain several species each.

---

* From the easily accessible literature, this group has been the simplest with which to commence laboratory study among the parasitic fungi. Its species are consequently the best known of any of our parasitic fungi.

† This appears to be the original orthography. The genus is often known as *Erysiphe*, and the family as Erysiphaceae.

The family Perisporiaceae contains some twenty genera and the Microthyriaceae as many more.

LITERATURE.

Lindau. Die natürlichen Pflanzenfamilien, 1^1: 325–343.

Saccardo. Sylloge Fungorum, 1: 1–87; 2: 658–671; 9: 364–442, 1053–1072; 11: 252–271, 379–382.

Leville. Organisation et disposition méthodique des espèces qui component les genre érysiphé. Ann. Sc. Nat. III. 15: 109–179. *Pl. 6–11.* 1851.

Tulasne. Selecta Fungorum Carpologia. 1: 1861.

Burrill & Earle. The parasitic Fungi of Illinois. Part II. Bull. Ill. State Lab. Nat. Hist. 2: 387–432. 1887.

Ellis & Everhart. North American Pyrenomycetes, 2–58. 1892. (The Erysiphaceae are by Burrill.)

Harper. Die Entwickelung des Perithecium bei Sphaerotheca Castagnei. Berichte deutsch. bot. Gesell. 13: 475–481. *Pl. 39.* 1895.

Martin. Synopsis of the North American Species of Asterina, Dimerosporium and Meliola. Jour. Myc. 1: 133–139, 145–148. 1885.

Order 6. HYPOCREALES.

The Hypocreales are for the greater part light colored fungi of various habits, some parasitic on the higher plants, others parasitic on fleshy fungi, others still on insects, while numerous species are saprophytic. Their color ranges in different species from white to yellow and on to purple, scarlet and cinnabar red. A few are plain brown but none of them are really black, a somewhat artificial distinction that separates this order from the Sphaeriales.

Among the members of this group one of the most common is known as ergot. This fungus produces enlarged hard bodies in the kernels of rye and other grasses; these are known as sclerotia[*] and in this condition they pass the winter season. In the spring the mycelium of the sclerotium becomes active and produces a series

---

[*] Sclerotia are produced on a number of species of fungi belonging to widely different groups. They are receptacles of nutrition in a dormant condition intended to carry the organism over an unfavorable period.

of stalked stromata with bright colored heads bearing embedded perithecia. The perithecia are pear-shaped and perforated by an ostiole at the narrow end ; the spores are filiform. When these germinate in the pistils of the rye flowers they produce a mass of mycelium that fills every part of the kernel and forms a white covering from which conidia are produced ; this is known as the *sphacelia* stage ; with these a saccharine fluid is developed which is attractive to bees ; in this way the fungus is spread from one flower to another, the bees carrying the conidia which are mixed with this fluid, the conidia soon germinating and producing new centres for the ergot. The sclerotium then forms in the midst of the mycelium-infested kernel and reaches its dormant condition by the time the grain is ripe.

From the ergot of rye a powerful poisonous drug is produced which is frequently used in medicine. A number of our common grasses are affected with ergot but the stromatic stage has been little studied. It is supposed that cattle feeding on the ergot will become poisoned from its effects, the disease being known as ergotism.

Cordyceps is a genus of fungi somewhat allied to the last, but instead of growing from sclerotia it becomes parasitic either upon insects or upon truffle-like subterranean fungi (*Elaphomyces*). The most common form on insects is a club-like body an inch or more long of a cinnabar red color which attacks the pupae of various moths buried beneath dead leaves, the club rising above the surface. Other species attack the larvae of beetles and a minute one is found on scale insects. In New Zealand a similar fungus attacks living caterpillars which carry the fungus on their backs for a considerable time, before it results in their death. The forms growing on truffles are somewhat fleshy structures with a conical stroma which pushes up above the surface of the ground. The American species of this group have never been thoroughly studied.

A number of other interesting genera commonly occur. One (*Epichloe typhina*) forms a whitish or yellowish covering on the culms of various grasses causing them to appear like miniature spikes of cat-tails ; another (*Hypocrella*) forms tubercles on the stems of cane (*Arundinaria*) ; the members of a third genus (*Hypomyces*) are parasitic on various forms of the higher fleshy, corky or gelatinous fungi, one of which attacks species of *Lactarius* transforming the gills and causing the surface to appear a bright reddish orange color ; several other species of various colors are found on *Russula ;* still another genus (*Nectria*) appears in the form of small or minute red tubercles on dead branches or trunks. In all, the single family Hypocreaceae* contains some sixty genera and over 800 species, one fourth perhaps of which are from the United States.

LITERATURE.

Saccardo. Sylloge Fungorum, 2 : 447–587 ; 9 : 941 1004 ; 11 : 354–368.

Lindau. Die natürlichen Pflanzenfamilien, 1¹: 343–372.

Tulasne. Selecta Fungorum Carpologia, 3 : 1865.

Tulasne. Mémoire sur l'ergot des Glumacées. Ann. Sc. Nat. III. 20: 1–56. *Pl. 1–4.* 1853.

Massee. A Revision of the Genus Cordyceps. Ann. Bot. 9: 1–44. *Pl. 1–2.* 1895.

Ellis & Everhart. Synopsis of the North American Hypocreaceae. Jour. Mycol. 2: 28–31, 49–51, 61–69, 73–80, 97–99, 109–111, 121–125, 133–137. 1886 ; 3 : 1–6. 1887.

———— Additions to Hypocreaceae. Jour. Myc. 3: 113–116. 1887.

———— The North American Pyrenomycetes, 58–122. 1892.

Order 7. DOTHIDEALES.

This order, consisting of a single family Dothideaceae with some 25 genera and 400 species, is made up of species growing mostly on dead or dying plant tissues, the perithecia being buried in a black or blackish stroma, the wall of the perithecium being indistinct and fused with the stromatic substance. Nearly half the species belong to the genus *Phyllachora*, very common on the leaves of grasses and other plants and forming lines and patches superficially placed like those of the black rusts. While most of the species are saprophytic, a few are genuine parasites ; among the most notorious is the black knot of the plum and cherry which well represents the stromatic condition. This fungus ap-

---

*Some mycologists regard this order as made up of several well-marked families. Lindau, *loc. cit*, arranges the genera in six tribes; some are disposed to regard these as families.

pears as a destructive disease in many parts of the country occasionally destroying entire orchards. The mycelium of the fungus spreads itself within the interior of the young branches and manifests itself by the formation of elongate gray-brown or blackish deformities which give the name to the fungus. These sometimes attain a length of five or six inches. In the spring these are covered with a velvety surface which is made up of a series of short thread-like hyphae bearing conidia; these conidia serve to extend the infection to other parts of the tree where they are carried. Later the knots are covered with rounded bodies which are the projecting portions of the perithecia mostly imbedded in the stromata; these contain the ascospores which mature late in the winter and escape by a pore at the upper end of the asci. Besides conidia and ascospores two other reproductive bodies have been found but their functions are little known.

LITERATURE.

The literature relating to the Dothideales is mostly associated with that of the next order.

Lindau. Die natürlichen Pflanzenfamilien, 1^1: 372-383.

Winter. Rabenhorst's Kryptogamen Flora Deutschlands, u. s. w. 1^2: 893-918.

Saccardo. Sylloge Fungorum, 2: 588-657; 9: 1004-1053; 11: 368-379.

Ellis & Everhart. The North American Pyrenomycetes, 596-621. 1892.

Farlow. The black knot. Bull. Bussey Inst. 440-454. *Pl.* *4-6*. 1876.

Order 8. SPHAERIALES.

The fungi belonging to this order are probably the most numerous of all the groups, there being nearly or quite two thousand species known from our own country alone tho many of them are known very imperfectly. They range in habit from leaf parasites to terrestrial forms tho the great majority grow on wood or other vegetable stems. Except in a few families like the Chaetomiaceae the mycelium is chiefly confined to the substratum.

Some species possess no stroma and the perithecia are either attached to a membranous subiculum or are entirely separate from each other. A common saprophytic form of this type is seen in the black conic papillae often found on the inner surface of the halves of old peach stones long exposed to the weather; these are the scattering perithecia of a black fungus (*Caryospora putaminum*, *Pl. 4, f. 15, 16*) which contain the large peculiar two-celled spores. A common parasitic form of the same type is seen in late summer on the leaves of hazel; in this species (*Mamiana coryli*) the black pear-shaped perithecia are separately located on definite areas on the under surface of the leaf.

In other species the perithecia may be united in a somewhat woody stroma as in the numerous black or red brown tubercles of *Hypoxylon* everywhere common on beech, birch, alder, oak and many other woods; other forms are seen in the smaller tubercles erumpent through the bark of twigs and branches of woody plants or in the persistent half woody stems of herbaceous plants.

In some genera the mouth (ostiole) of the perithecia is flat, while in others the perithecia become conic or even rostrate (*Pl. 4, f. 17*). In many cases the formation of conidia precedes the development of the ascospores, and in a few of these there is a superficial resemblance to fungi of the next class which has deceived even botanists familiar with these plants.

In a few cases where there is no definite stroma a black layer is formed about the mouth of the sunken perithecia in the shape of a rounded disc. This is known as the *clypeus*, and characterizes an entire family.

The families* may be separated as follows:

1. Perithecia free, either without a stroma, partly sunken in a loose mass of mycelium, or sessile above an imperfect stroma. 2.
    Perithecia sunken in the substratum, without a stroma, rarely united above by a black tissue (clypeus). 9.
    Perithecia fully imbedded in a stroma, the mouths only projecting, or becoming free by the breaking away of the outer stromatic layers. 13.

---

*We have omitted generic synopses in this numerous order for several reasons. (1) Because of their great number, which would unduly extend the limits of a small work. (2) Because of the difficulty for a beginner to distinguish closely allied genera, and (3) Because any one sufficiently interested in the species will need the literature cited at the close of the section, in which descriptions of the species as well as synopses of the genera may be found. In spite of all that has been written, however, many of our species are imperfectly known, even in their ascosporic stages, and there is scarcely a genus that is not in crying need of a revision.

2. Walls of the perithecia thin and membranous; asci soon disappearing. 3.
   Walls of the perithecia coriaceous or carbonaceous. 4.

3. Perithecia always superficial, with copious tufts of hair at the mouth.
   Chaetomiaceae. (2 gen.)*
   Perithecia usually sunken, with short hairs about the mouth or none;
   growing on manure. **Sordariaceae.** (7 gen.)

4. Perithecia either entirely free, or at most with the base slightly sunken
   in the substratum or stromatic layer. 5.
   Perithecia more or less deeply sunken in the substratum at base, free
   above. 8.

5. Stroma wanting or merely thread-like or tomentose. 6.
   Stroma present. 7.

6. Mouths of the perithecia mostly in the form of short papillae.
   Sphaeriaceae. (25 gen.)
   Mouths of the perithecia more or less elongate, often hair-like.
   Ceratostomaceae. (8 gen.)

7. Stroma mostly well developed with an indefinite border; perithecia in
   close irregular masses, never flask-like nor funnel-like at the apex.
   Cucurbitariaceae. (9 gen.)
   Stroma small, sharply bordered; perithecia in rows or in regular rounded
   masses, flask-shaped, with funnel-formed mouths.
   Coryneliaceae. (3 gen.)

8. Mouths of the perithecia circular in outline.
   Amphisphaeriaceae. (9 gen.)
   Mouths of the perithecia laterally compressed.
   Lophiostomaceae. (8 gen.)

9. Asci usually thickened at the apex, breaking open by a pore; mouths
   of the perithecia mostly beaked (rarely only rounded). 10.
   Asci not thickened at the apex, mostly projecting at maturity. 11.

10. Perithecia with a clypeus. **Clypeosphaeriaceae.** (8 gen.)
    Perithecia without a clypeus. **Gnomoniaceae.** (10 gen.)

11. Walls of the perithecia thin, coriaceous; mouth mostly short or
    plane. 12.
    Walls of the perithecia carbonaceous or thick coriaceous; spores large,
    mostly enveloped in gelatine. **Massariaceae.** (10 gen.)

12. Asci clinging together in fascicles, without paraphyses.
    Mycosphaerellaceae. (12 gen.)
    Asci not fascicled; paraphyses present. **Pleosporaceae.** (23 gen.)

---

* To give some idea of the relative size of the families we have added
the number of genera in each in parenthesis.

13. Stroma fused with the substratum. 14.
    Stroma formed almost wholly of hardened fungal hyphae. 15.

14. Conidia developed in pycnidia. **Valsaceae.** (10 gen.)
    Conidia developed from a flattened surface.
    Melanconidaceae. (9 gen.)

15. Spores small, cylindric, 1-celled, mostly curved, hyaline or yellowish-brown. **Diatrypaceae.** (8 gen.)
    Spores rather large, 1–many-celled, hyaline or brown; conidia mostly in cavities of the stroma. **Melagrammataceae.** (9 gen.)
    Spores 1-celled (rarely 2-celled), blackish brown; conidia developed on the upper surface of the young stromata.
    Xylariaceae. (14 gen.)

It will become apparent in using the above synopsis that many of the characters used to distinguish some of the families gradually shade into each other. This arises in part from the existence of numerous intermediate or connecting forms in this order, and in part from our lack of positive information relative to many of the genera. In the determination of genera it is sometimes simpler to follow the artificial arrangement of Saccardo in whose *Sylloge Fungorum* natural alliances are largely passed over and genera are arranged according to their spore forms and colors.

With the exception of the Xylariaceae most of the plants of the entire order are comparatively inconspicuous. Besides the species of *Hypoxylon*, noted above, the family Xylariaceae contains *Daldinia* with three or four species which form blackish or brownish masses, often as large as butternuts, of an ovoid or spherical form. *D. concentrica* shows a series of concentric brown layers within the stroma, while *D. vernicosa* when young shows a similar series of layers, the main portions of each being composed of whitish membranous septa. Both are comparatively common. Other genera are *Ustulina*, which forms flattish, irregular, crust-like masses growing commonly on maples; *Nummularia* forms more regular crust-like, smooth masses bursting through the young bark of sapling oaks or sometimes occurs on the branches of larger trees; *Xylaria* forms simple or branching club-like masses growing on logs, or at the bases of trees or stumps attached to the wood beneath the surface of the ground; *X. polymorpha* (*Pl. 4. f. 11, 12*) is a common species often growing in large clusters; *Poronia* contains curious stalked blackish bodies growing on manure mainly in the Southern States.

LITERATURE.

Saccardo. Sylloge Fungorum, 1: 88-754; 2: 1-446; 672-720; 9: 442-940; 11: 271-353, 382-385.

Lindau. Die natürlichen Pflanzenfamilien, 1^1: 384-491.

Winter & Rehm. Rabenhorst's Kryptogamenflora Deutschland, u. s. w. 1^2: 152-893.

Tulasne. Selecta Fungorum Carpologia, 2:

Ellis & Everhart. North American Pyrenomycetes, 122-673. 1892.

Berlese. Icones Fungorum (current). 1891 —.

Ellis & Everhart. Synopsis of the North American species of Xylaria and Poronia. Jour. Mycol. 3: 97-102, 109-113. 1887.

——— Synopsis of the North American species of Hypoxylon and Nummularia. Jour. Mycol. 4: 38-44, 66-70, 85-93, 109-113. 1888; 5: 19-23. 1889.

Order 9. LABOULBENIALES.

The Laboulbeniales form a group of fungi not closely related to any other orders. They are parasitic on various species of insects, more commonly beetles, and are exceedingly minute. From their apparent sexual method of reproduction and the semi-aquatic habits of some of the species they appear to find their nearest relations with some of the red algae. They are connected with their host by means of a dark-colored horny piece which serves both as a hold-fast and as an organ of nutrition. The fungus consists of a somewhat club-shaped receptacle made up of a few cells and bearing at its end one or more perithecia, in which the asci are developed by successive sprouting from basal cells. The systematic relations of the order are not well known. They have been made a subject of special study by Dr. Roland Thaxter, of Cambridge, Mass., who has brought to light many more forms from this country than are known from all the rest of the world. Nearly 150 species are now known, mainly through his studies. Of the twenty-eight recognized genera Dr. Thaxter has proposed twenty-four.

LITERATURE.

Thaxter. On some North American species of Laboulbeniaceae. Proc. Amer. Acad. Arts and Sci. **26** : 5–14. 1890.

────── Supplementary note on North American Laboulbeniaceae. *Ibid.* 261–270. 1891.

────── Further additions to the North American species of Laboulbeniaceae. *Ibid.* **27** : 29–45. 1892.

────── New species of Laboulbeniaceae from various localities. *Ibid.* **28** : 156–188. 1893.

────── New genera and species of Laboulbeniaceae with a synopsis of the known species. *Ibid.* **29** : 92–111. 1894.

────── Notes on Laboulbeniaceae, with descriptions of new species. *Ibid.* **30** : 467–481. 1895.

All the above are short papers, simply preliminary to the next elaborate monograph.

────── Contribution towards a Monograph of the Laboulbeniaceae. Mem. Amer. Acad. **12** : 187–429. *IV.* 1–26. 1896.

Order 10. TUBERALES.

The members of this order are subterranean fungi resembling tubers. Some of the species known as truffles are highly prized for food and command high prices in European markets, two average sized canned truffles often retailing for three francs. The truffle varies from the size of an acorn to the size of a fist and has a warty appearance on the outside. The asci are formed on the interior of the fungus. Little is known of the method of spore dissemination. The edible species have not yet been found in America, but several smaller and unimportant forms have occasionally been found. The truffles are not to be confounded with the various forms of subterranean puff-balls which are comparatively common in the Southern States, nor with certain of the subterranean Aspergillales which resemble them even more closely since they produce their spores in asci. In Southern Europe the true truffles are hunted either by dogs or pigs trained for the purpose.

The two families of Tuberales are distinguished as follows :

Ascocarps formed of several labyrinthine passages opening outward at maturity. **Tuberaceae.**

Ascocarps with a single or several closed cavities not opening outward at maturity. **Balsamiaceae.**

Of the Tuberaceae, *Pseudhydnotrya Harknessi* is reported from California, and *Tuber macrosporum* from the Eastern States. Among the Balsamiaceae, *Geopora Cooperi* is known from California.

LITERATURE.

Fischer, E. Die natürlichen Pflanzenfamilien, 1^1 : 278–290.

——— Rabenhorst's Kryptogamen flora Deutschland u. s. w. 1^5 : 1–131.

Saccardo. Sylloge Fungorum, **8** : 872–908 ; **10** : 80–83, **11** : 442–445.

Chatin. La truffe. Paris. 1892.

Order 11. HYSTERIALES.

The Hysteriales are represented by small species, elongate or often boat-shaped, possessing a covering to the ascoma, which ruptures at length, opening by a longitudinal slit. Some of the species are parasitic on leaves and resemble scale-insects. The members of this order are more tropical in their distribution and many of them may be expected in the gulf region of America. Quite a number, however, are found in the northern states, mostly occurring as saprophytes on dry decorticated twigs and rotting wood. *Dichaena faginea* is common everywhere on beech trees, forming large blackish blotches on the gray bark. The families may be distinguished as follows :

1. Ascocarps immersed ; walls of the ascocarps connate with the membranous covering. **Hypodermataceae.**
Ascocarps immersed at first, erumpent at maturity ; walls free, membranous or carbonaceous. 2.
Ascocarps free ; walls carbonaceous or membranous. 3.
2. Walls membranous or coriaceous, black. **Dichaenaceae.**
Walls thick, almost corky, gray or black. **Ostropaceae.**
3. Walls carbonaceous, black ; shield round, oval or more commonly linear. **Hysteriaceae.**
Walls membranous or horny, brown ; ascocarps vertical, clavate.
Acrospermaceae.

Of the above families the Dichaenaceae and the Acrospermaceae, each contain a single genus from which the families are respectively named. The family Ostropaceae contains two genera.

The remaining two families, Hypodermataceae with nine genera and the Hysteriaceae, best known of all, with fourteen genera contain the majority of the species from America and elsewhere.

LITERATURE.

Saccardo. Sylloge Fungorum, 2: 721-813; 9: 1094-1129; 11: 385-390.

Rehm. Rabenhorst's Kryptogamen Flora Deutschlands, u. s. w. 1^a: 1-56.

Ellis & Everhart. North American Pyrenomycetes, 673-727. 1892.

Duby. Memoire sur la Tribu des Hysterinées. Mém. Soc. Phys. et d'Hist. Nat. Genève, 16: 15-70. *pl. 1, 2.* 1861.

Lindau. Die natürlichen Pflanzenfamilien 1^1: 265-278.

Order 12. PHACIDIALES.

The Phacidiales are partly saprophytic and partly parasitic plants in which the ascoma is usually roundish or stellate and remains enclosed for a long time in a tough covering which becomes torn at maturity of the spores. The order is composed of three families which may be distinguished as follows :

1. Ascocarps soft, fleshy, bright-colored ; disc mostly bright-colored, surrounded by the lobes of the ascocarp (saprophytic) **Stictidaceae.**
   Ascocarp leathery or carbonaceous, always black. 2.
2. Ascocarps at first sunken, later strongly erumpent, hypothecium thick (saprophytic). **Tryblidiaceae.**
   Ascocarps remaining sunken in the substratum ; hypothecium thin, poorly developed (parasitic or saprophytic). **Phacidiaceae.**

All the above families are of considerable size, the Stictidaceae having twenty-two genera, the Tryblidiaceae six, and the Phacidiaceae seventeen. Among the parasitic members of the last named family, *Rhytisma* is one of the most familiar genera, species occurring on the leaves of maple, holly, willow and andromeda, forming raised black blotches ; one of the two species on maple leaves is often abundant and conspicuous. A species of *Phacidium* causes the brown spots frequently seen on the leaves of sweet clover. *Trochila* also contains some leaf-inhabiting parasites.

The American genera and species of the order have never been systematically studied.

LITERATURE.

Saccardo. Sylloge Fungorum, 8: 705–811; 10: 48–67; 11: 431–435.

Lindau. Die natürlichen Pflanzenfamilien, 1¹: 243–265.

Rehm. Rabenhorst's Kryptogamen Flora Deutschland u. s. w. 1³: 59–212.

Order 13. PEZIZALES.

The Pezizales or cup fungi form a very extensive group of mostly saprophytic plants. They are typically disc-shaped or cup-shaped and when young are closed or nearly so, opening as they mature. They vary in size from minute species scarcely visible to the naked eye to large fleshy forms three or four inches in diameter. A few species possess a stalk of considerable length but the greater number are either sessile or short-stalked. (*Pl. 4. f. 9, 14.*) Most grow either on the ground or on various decaying vegetable substances. A few forms are parasitic on living plants.

In substance the ascocarp may be fleshy, waxy, leathery, horny or in a few cases gelatinous. The paraphyses may be either free or united into a sort of covering to the ascoma known as the epithecium. The cup or disc-shaped ascoma is often separable into two layers, one known as the hypothecium which contains the stratum of asci, and the other the peridium which forms the outer portion. In many cases these two layers are not clearly distinguishable. In a number of instances the spores are ejected from the asci with an explosive force and in an entire family the asci themselves are projected from their bed often with an audible explosion.

Of the ten families of Pezizales, nine * are represented in North America. They may be distinguished by the following synoptic table :

---

* The Cyttariaceae of the southern hemisphere are curious compound ascomata arranged in a globular stroma. The fungus grows attached to bushes.

1. Ascocarps free, solitary or cespitose. 2.
   Ascocarps affixed to the ends of the branches of a cord-like stroma.
                           Family 9. Cordieritidaceae.*
2. Ascocarps fleshy or waxy, rarely gelatinous; ends of paraphyses free. 3.
   Ascocarps leathery, horny or cartilaginous; ends of paraphyses united into an epithecium. 7.
3. Peridium and hypothecium without distinct line of junction. 4.
   Peridium forming a more or less differentiated membrane. 6.
4. Ascomata open and convex from the beginning; peridium wanting or poorly developed.   Family 1. Pyronemaceae.†
   Ascomata concave at first; peridium fleshy. 5.
5. Asci forming a uniform stratum at maturity. Family 2. Pezizaceae.
   Asci projected from the ascoma at maturity. Family 3. Ascobolaceae.
6. Peridium formed of elongate parallel pseudo-parenchyma with clear and thin-walled cells.   Family 4. Helotiaceae.
   Peridium firm, of roundish or angular pseudo-parenchyma, with mostly thick and dark-colored cell walls.   Family 5. Mollisiaceae.
7. Peridium wanting or poorly developed. Family 6. Celidiaceae.
   Peridium well developed, mostly leathery or horny. 8.
8. Ascocarps free from the beginning, dish or plate-shaped, never enclosed by a membrane.   Family 7. Patellariaceae.
   Ascocarps at first embedded in the matrix, then erumpent, urceolate or cup-shaped, at first often enclosed in a membrane which disappears later.   Family 8. Cenangiaceae.

Family 2   Pezizaceae.‡

Our genera of Pezizaceae which contain many of the more conspicuous of the fleshy cup fungi may be separated as follows :

1. Spores globose. 2.
   Spores ellipsoid, blunt or more rarely acute. 4.

---

* A single species of *Cordieritis* is found in the Southern States.

† Of the four genera composing this family there are with us only four species of *Pyronema*.

‡ Since there is no available English synopsis of the genera of Pezizales and as the American species have hitherto been neglected and need careful systematic study in the field and laboratory, we give rather more space to them than to some groups in which manuals are accessible. It is hoped that more attention will thus be called toward this inviting group.

2. Ascomata hairy outside. 3.
  Ascomata smooth outside. PLICARIELLA.
3. Hairs long, sharp-pointed; ascoma bright colored. SPHAEROSPORA.
  Hairs fine, short; ascoma dark colored. PSEUDOPLECTANIA.
4. Ascomata hairy. 5.
  Ascomata smooth. 6.
5. Ascomata circular. LACHNEA.
  Ascomata stellately lobed, partly underground. SARCOSPHAERA.
6. Ascomata regularly saucer-shaped or cup-shaped, circular. PEZIZA.*
  Ascomata stellately lobed partly under ground. SARCOSPHAERA.
  Ascomata irregular, dimidiate or ear-shaped. OTIDEA.

Of the above genera, *Lachnea* and *Peziza* especially contain many species.

Family 3. Ascobolaceae.

The Ascobolaceae are peculiar in their habit of discharging their asci so as to more widely disseminate their spores. They are commonly found growing on old cow dung and similar locations. Our genera can be quite readily distinguished:

1. Spores hyaline, spherical. CUBONIA.
  Spores hyaline, ellipsoid. 2.
  Spore sat length violet or brownish. 5.

---

*The large genus *Peziza* is made up of several groups of species (subgenera), which are often regarded sa genera. They may be separated as follows:

1. Juice colored and milky, exuding when wounded. GALACTINIA.
  Juice watery, colorless. 2.
2. Asci blue when treated with iodine. 3.
  Asci not becoming blue when treated with iodine. 4.
3. Ascomata sessile. PLICARIA.
  Ascomata on a stalk. TAZETTA.
4. Spores smooth or irregularly tuberculate or warty. 5.
  Spores at length covered with reticulations. ALEURIA.
5. Ascomata entirely sessile. HUMARIA.
  Ascomata more or less stalked. 6.
6. Stalk short, thick, smooth 7.
  Stalk short, fluted or grooved. ACETABULA.
  Stalk long, thin, mealy or rough tuberculate outside. MACROPODIA.
7. Ascomata goblet-shaped or cup-shaped. GEOPYXIS.
  Ascomata at length becoming spread out like a disc. DISCINA.

2. Peridium developed. 3.
   Peridium wanting. ZUKALINA.*
3. Asci 8-spored. 4.
   Asci 16-∞-spored. RHYPAROBIUS.†
4. Ascomata hairy. LASIOBOLUS.
   Ascomata smooth. ASCOPHANUS.
5. Spores spherical. BOUDIERA.
   Spores ellipsoid, joined in a ball within the ascus. SARCOBOLUS.
   Spores ellipsoid, free. ASCOBOLUS.

Family 4. **Helotiaceae.**

This family differs from the Pezizaceae mainly in the possession of a distinct peridial layer, a condition which is usually easily determined by making a vertical section through the cup. This condition renders the fungus of a tougher consistency and hence less perishable and more persistent. The bright scarlet cups that are found on sticks in early spring (*Sarcoscypha coccinea*), the equally handsome hairy goblet-like cups of midsummer (*Sarcoscypha floccosa*) and the small bright egg-yellow disc-like fungi (*Helotium citrinum*) that are common all summer under logs, chips and the like, are representative species. One species, *Chlorosplenium aeruginosum*, is peculiar in its color, the fungus, mycelium, ascocarp and all being a verdigris green; the ascocarps are not very common but the mycelium is very common in decaying wood staining the whole woody substance a peculiar copper-green color. The American genera of this rather large family can be distinguished by the following synopsis:

1. Ascocarps waxy or fleshy-waxy, thick or membranous 2.
   Ascocarps gelatinous gristly, horny when dry. 21.
2. Ascocarps fleshy-waxy, brittle when fresh, leathery when dry. 3.
   Ascocarps waxy, thick, tough or membranous. 7.
3. Ascocarps felty, hairy externally. SARCOSCYPHA.
   Ascocarps covered with bristle-like hairs externally. PILOCRATERA.
   Ascocarps externally naked. 4.

---

*The genus *Glœopeziza* from Austria is distinguished from this by having the disc covered with a gelatinous layer when young.

†The genus *Streptotheca* of Europe is distinguished from this by the asci being provided with a ring below the apex.

4. Ascocarps springing from a sclerotium.   SCLEROTINIA.
    Ascocarps not springing from a sclerotium.   5.

5. Spores one-celled.   6.
    Spores at length 2-4-celled.   RUTSTROEMIA.

6. Substratum colored green.   CHLOROSPLENIUM.
    Substratum uncolored.   CIBORIA.*

7. Ascocarps hairy externally.   8.
    Ascocarps naked.   15.

8. Ascomata resting on an extended arachnoid mycelium.   9.
    Ascomata without arachnoid mycelium.   10.

9. Spores remaining one-celled.   ERIOPEZIZA.
    Spores becoming several celled.   ARACHNOPEZIZA.

10. Spores globose.   LACHNELLULA.
    Spores ellipsoid or elongate.   11.

11. Disc surrounded with black hairs.   DESMAZIERELLA.
    Disc smooth.   12.

12. Paraphyses obtuse at the apex.   13.
    Paraphyses lancet-shaped at the apex.   14.

13. Walls of ascoma delicate; spores mostly one-celled, sometimes two-celled at maturity.   DASYSCYPHA.
    Walls of ascoma thick; spores two-celled at maturity.   LACHNELLA.

14. Spores remaining one-celled.   LACHNUM.
    Spores at length several-celled.   ERINELLA.

15. Spores globose.   PITYA.
    Spores ellipsoid or fusiform.   16.
    Spores filiform.   20.

16. Spores remaining one-celled.   17.
    Spores at length two-four-celled.   18.

17. Border of disc smooth.   HYMENOSCYPHA. †
    Border of disc toothed.   CYATHICULA.

18. Ascocarps sessile, rarely compressed at base.   BELONIUM.
    Ascocarps stalked, or at least compressed like a stalk.   9.

19. Walls of ascoma waxy; stem short and delicate.   BELONIOSCYPHA.
    Walls of ascoma waxy, thick; stem thick.   HELOTIUM.

---

* Some species of *Pilocratera* may also be sought here.
† Some forms of *Helotium* may be sought here.

20. Ascocarps sessile. GORGONICEPS.
    Ascocarps stalked. POCILLUM.
21. Spores remaining one-celled 22.
    Spores at length many-celled. CORYNE.
22. Ascocarps globose, at first sessile, at length cup-shaped with a short stalk, small (1 mm. wide). STAMNARIA.
    Ascocarps stalked from the first, clavate; disc at length often saucer-shaped, larger. OMBROPHILA.

While many of the above genera are small, *Sarcoscypha* and *Erinella* have about twenty species each, *Ombrophila* thirty, *Sclerotinia* and *Lachnella* each forty, *Dasyscypha* and *Lachnum* each one hundred and fifty and *Hymenoscypha* and *Helotium* each two hundred. These include species from the whole world. The species from our own country are very imperfectly known.

Family 5. Mollisiaceae.

The numerous species of this family are mostly inconspicuous cup fungi growing on stems, fallen leaves or, as in some species of *Pseudopeziza* and *Pyrenopeziza*, parasitic on living plants; *Pseudopeziza trifolii* on the leaves of clover is a common and injurious parasite. The genera can be distinguished as follows:

1. Ascocarps fleshy, waxy or rarely membranous. 2.
   Ascocarps gelatinous gristly, horny when dry. 11.
2. Ascocarps sunken in the substratum at first, at length erumpent. 3.
   Ascocarps not sunken in the substratum. 7.
3. Ascocarps bright colored, only slightly erumpent. 4.
   Ascocarps dark colored, at length strongly erumpent. 5.
4. Spores ellipsoid or elongate, rounded, one-celled. PSEUDOPEZIZA.
   Spores becoming many celled. FABRAEA.
5. Spores ellipsoid or fusiform one-celled. 6.
   Spores many-celled by transverse septa. BELONIELLA.
6. Ascocarps bristly externally and on the margin. PIROTTAEA.
   Ascocarps externally smooth, the margin at most merely shredded. PYRENOPEZIZA.
7. Ascocarps seated on an often radiate mycelium. 8.
   Ascocarps not seated on a visible mycelium. 9.
8. Spores elongate, often fusiform, one-celled. TAPESIA.
   Spores filiform, many-celled. TRICHOBELONIUM.

9. Spores remaining one-celled. 10.
   Spores becoming 2-celled. NIPTERA.
   Spores elongate-fusiform, 4-∞-celled. BELONIDIUM.
   Spores filiform, ∞-celled. BELONOPSIS.
10. Spores spherical. MOLLISIELLA.
    Spores elongate. MOLLISIA.
11. Spores remaining one-celled. ORBILIA.
    Spores finally 2-4-celled. CALLORIA.

Family 6. Celidiaceae.

This family contains a few inconspicuous genera mostly growing on lichens, rarely on wood or bark. Tho several of the genera have not yet been reported from North America, a synopsis is here given since any of them are likely to be found here.

1. Spores one-celled. 2.
   Spores 2-celled. 3.
   Spores 4-6-celled; growing on lichens. CELIDIUM.
2. Growing on wood or bark. AGYRIUM.
   Growing on lichens. PHACOPSIS.
3. Growing on wood or bark. LECIDEOPSIS.
   Growing on lichens. CONIDA.

Family 7. Patellariaceae.

1. Walls of the ascoma thin; hypothecium only slightly developed. 2.
   Walls of ascoma thickened; hypothecium well developed. 3.
2. Spores hyaline, 1-celled or at maturity, sometimes 2-celled. PATELLEA.
   Spores hyaline, 4-6-celled. DURELLA.
   Spores brown, 2-celled. CALDESIA.
3. Asci 8-spored. 4.
   Asci 16-spored; spores 2-celled. RAVENELULA.
   Asci many-spored. 14.
4. Spores remaining 1-celled at maturity. 5.
   Spores becoming 2-celled at maturity. 7.
   Spores elongate, needle-shaped or filiform, 4-many-celled. 11.
5. Paraphyses not broadened above, wavy. STARBAECKIA.
   Paraphyses enlarged upwards (clavate). 6.
6. Ascocarps superficial from the beginning (saprophytic). PATINELLA.
   Ascocarps immersed at first, then erumpent (parasitic). NESOLECHIA.

7. Ascocarps naked. 8.
   Ascocarps hairy, parasitic on living leaves. JOHANSONIA.
8. Spores hyaline at maturity. SCUTULA.
   Spores at first hyaline, then brown or brownish. 9.
9. Ascocarps superficial or rarely slightly immersed (saprophytic). 10.
   Ascocarps at first immersed, then erumpent (parasitic). ABROTHALLUS.
10. Disc round. KARSCHIA.
    Disc elongate or irregular. MELASPILEA.
    Disc linear or sometimes stellately branched. HYSTEROPATELLA.
11. Spores not breaking up into single cells in the asci. 12.
    Spores filiform, many-celled, breaking up into single cells in the asci. BACTROSPORA.
12. Spores ellipsoid, mostly 4- (rarely 6-8-) celled, hyaline, then brown (mostly parasitic). LECIOGRAPHA.
    Spores fusiform, 4- or more celled, hyaline (saprophytic). PATELLARIA.
    Spores filiform, elongate. 13.
13. Ascocarps sessile; spores bacillate, 4-6-celled. PRAGMOPORA.
    Ascocarps sessile; spores filiform, many-celled. SCUTULARIA.
    Ascocarps top-shaped, stalked. LAHMIA.
14. Spores roundish, 1-celled. BIATORELLA.
    Spores elongate, 4-celled. BAGGEA.

The Patellariaceae are mostly inconspicuous saprophytic plants; the American species have never been systematically studied.

Family 8. Cenangiaceae.

1. Ascocarps coriaceous, corneous or waxy when fresh. 2.
   Ascocarps gelatinous when fresh. 8.
2. Ascocarps at first immersed, without a stroma. 3.
   Ascocarps springing from a more or less developed stroma. 7.
3. Spores one-celled. 4.
   Spores elongate, 2-4-celled. 5.
   Spores filiform, many-celled. GODRONIA.
4. Ascocarps externally bright colored, downy. VELUTARIA.
   Ascocarps externally dark, smooth; spores hyaline. CENANGIUM.
   Ascocarps externally dark, downy; spores colored. SCHWEINITZIA.
5. Spores hyaline, always 2-celled; ascocarps smooth. CENANGELLA.
   Spores hyaline, 2-4-celled; ascocarps downy externally. CRUMENULA.
   Spores at length brown or blackish. 6.

6. Disc elongate with a thick rim. TRYBLIDIELLA.
   Disc roundish with a thin rim ; spores 2-celled. PSEUDOTRYBLIDIUM.
   Disc roundish with an involute rim ; spores 4-celled. RHYTIDOPEZIZA.

7. Spores 8, not sprouting in the ascus. DERMATEA.
   Spores sprouting in the asci which become filled with small conidia.
   TYMPANIS.

8. Ascocarps sessile or stalked with smooth, saucer-shaped disc. 9.
   Ascocarps with convolute, tremelliform discs. 12.

9. Spores 1-celled, round. PULPARIA.
   Spores 1-celled, elongate. 10.
   Spores 2-celled. 11.
   Spores filiform. HOLWAYA.
   Spores muriform. SARCOMYCES.

10. Ascocarps soft, gelatinous inside, sessile, thin. BULGARIELLA.
    Ascocarps soft, gelatinous, stalked, thick. BULGARIA.
    Ascocarps watery gelatinous. SARCOSOMA.

11. Spores unequally 2-celled, rounded at the ends (parasitic on algae growing on bryophytes). PARYPHEDRIA.
    Spores elongate, acute at the ends (growing on wood). SOROKINA.

12. Spores 1-celled, hyaline. HAEMATOMYCES.
    Spores muriform, blackish. HAEMATOMYXA.

Species of *Bulgaria* are common on oak and chestnut forming top-shaped masses which are soft gelatinous when wet but shrivel into hard dry shapeless knobs. A species of *Dermatea* is common on stems of *Alnus* breaking out from underneath the bark. Some of the gelatinous species bear a superficial resemblance to certain of the true tremellines (Tremellales) but can be readily distinguished by the spores in asci.

LITERATURE.

Saccardo. Sylloge Fungorum, 8 : 53–646, 768–811 ; 10 : 3–44, 52–67 ; 11 : 393–427, 433–435.

Lindau. Die natürlichen Pflanzenfamilien, 1¹ : 173–273.

Rehm. Rabenhorst's Kryptogamen Flora Deutschlands, u. s. w. 1³ : 191–1134.

Cooke. Mycographia. *Pl. 1–113.* London, 1879.

Gillet. Les Champignons. Discomycetes. *Pl. 1–102.* Alençon, 1879 ; Suites. *Pl. 1–35.* 1890.

Phillips. Synopsis of the British Discomycetes. *Pl. 1–15.* London, 1877.

The only list of American species, now much out of date is :

Cooke. Synopsis of the Discomycetous Fungi of the United States. Bull. Buffalo Soc. Nat. Sci. 2 : 285–300; 3 : 21–37. 1875.

Order 14. HELVELLALES.

The Helvellales are fleshy fungi with an ascoma open from the earliest stage of its development. The asci are formed at the ends of hyphae which are variously interlaced to produce an ascocarp of a definite form. The asci form a definite layer and are usually mingled with sterile hyphae of variously modified forms (paraphyses).

Three families are known, all well represented in America as follows :

1. Ascocarps flat or arched, stemless ; asci opening by opercula.
   Rhizinaceae.
   Ascocarps formed of stem and ascoma, or columnar and stemless.   2.
2. Ascoma clavate or capitate ; asci opening by a terminal pore.
   Geoglossaceae.
   Ascoma conic or pileate ; asci opening by opercula.   **Helvellaceae.**

The Rhizinaceae contain two genera of rather unusual species and form a link with the cup-fungi of the preceding order. They may be separated as follows :

Ascoma fleshy, flattish, smooth beneath.   PSILOPEZIA.
Ascoma arched with root-like fibrils beneath.   RHIZINA.

The second family, the Geoglossaceae, are represented in America by nine genera, and form yellow, green or black club-like forms ranging from less than an inch to three inches in height. They are commonly terrestrial, growing in rich leaf mould, or often on decaying logs. Superficially they resemble some of simple club fungi (Clavariaceae) with which indeed the earlier mycologists united them. A section, however, readily discloses the spores borne in asci while in the Clavariaceae the spores are borne externally on basidia.

The American genera may be separated as follows :

1. Ascomata clavate or subcapitate, continuous with the stipe. 2.
   Ascomata flat; decurrent on both sides of the stipe; spores rod-like.
   SPATHULARIA.
   Ascoma capitate or hollow-discoid, usually with a free margin. 5.
2. Spores one-celled, colorless. 3.
   Spores 2–many-celled, with cross septa. 4.
3. Light colored, usually yellowish or light brown; ascomata sharply separated from the stipe. MITRULA.
   Bright or dark colored; ascomata not clearly separated from the stipe.
   MICROGLOSSUM.
4. Spores colorless. LEPTOGLOSSUM.
   Spores brown. GEOGLOSSUM.
5. Spores ellipsoid. 6.
   Spores elongate-filiform. 7.
6. Gelatinous-gristly. LEOTIA.
   Waxy. CUDONIELLA.
7. Fleshy; ascomata concave, hat shaped, the margin free, incurved.
   CUDONIA.
   Waxy; ascomata discoid above, the margin adnate to the stipe.
   VIBRISSEA.

Of the above genera we have one species in *Vibrissea*, two each in *Cudonia*, *Cudoniella*, and *Spathularia*, five in *Leotia* and six in *Mitrula* (*Pl. 4. f. 13*). The other genera are larger and may be divided into well marked sections. *Geoglossum* is represented by ten species arranged in two sections: § *Eugeoglossum* with smooth stipes, containing five species, and § *Trichoglossum* with hairy or bristly stipes, also containing five species.

Leptoglossum also contains two sections: § *Euleptoglossum*-blackish, containing two species, and § *Xanthoglossum*, yellowish or yellowish-brown, containing only *L. luteum*.

Microglossum is likewise formed of two sections: § *Eumicroglossum*, containing the dark colored (olive-green) species, *M. viride*, and § *Geomitrula*, containing the seven bright colored (yellowish or reddish) species. These were all united to *Mitrula* by Saccardo.

Of this family only *Leotia lubrica* is known to be edible. We have frequently seen this species growing in wet woods in Connecticut so abundant that several quarts could be gathered from an area of a few square rods.

The Helvellaceae contain the largest Ascomycetes known, some species of *Gyromitra* weighing over a pound, and forms of *Morchella* are occasionally a foot high. Among the most common genera are *Morchella* and *Gyromitra*, both of which are regarded as great delicacies in Europe and are quite generally eaten in this country. The morel (see frontispiece) is known in some parts of the country as "the spring mushroom," and in the upper Mississippi valley where it is usually very common in spring in woods, low ground, or about old stumps in orchards, it is more commonly known as "the mushroom" to the exclusion of other species. *Helvella* also contains several edible species. The six genera may be readily recognized by the following synopsis :

1. Ascomata with a distinct stalk. 2.
   Ascomata columnar, not stalked, the interior formed of several longitudinal chambers. UNDERWOODIA.
2. Ascomata conical or gyrose, hollow at least in the upper portion. 3.
   Ascomata campanulate or saddle-shaped, attached to the stipe at the middle. 4.
3. Pileus oval or conic, the upper surface consisting of deep pits formed by longitudinal and transverse ridges. MORCHELLA.
   Pileus irregular or lobed, the upper surface covered with gyrose wrinkles. GYROMITRA.
4. Pileus campanulate. VERPA.
   Pileus flat or arched, almost discoid. CIDARIS.
   Pileus lobed, irregular or saddle-shaped. HELVELLA.

Of the above genera *Cidaris* is known only by the single species decribed by Schweinitz, which has not since been found. *Underwoodia* also contains a single rare species only sparingly found in a single locality.*

Verpa contains two species and possibly a third representing two well marked sections : § *Ptychoverpa*, with thick, simple or forked, longitudinal ridges on the pileus, is represented by *V. bohemica*, and § *Euverpa* with a smooth pileus represented by *V. conica*, and a second species with a dark colored pileus that may be identical with *V. atro-alba* Fries.

---

* Kirkville, Onondaga County, New York. Only six plants were found in three different years. Schroeter placed this anomalous genus in the Rhizinaceae, but its affinities are more clearly in this family.

The remaining genera contain a number of species. *Helvella* is the largest genus, represented in this country by twelve species. These are divided into three somewhat natural groups according to the nature of the stipe. (1) Stipe thick, sulcate or furrowed. (2) Stipe thick, smooth. (3) Stipe slender, smooth (*i. e.*, not sulcate). The second contains only *H. monachella*, while the remaining species are about equally divided between the remaining sections.

Morchella contains the morels of which we have several species representing two distinct sections. (1) Those with the lower part of the pileus free and surrounding the stems (§ *Mitrophora*); and (2) those with pileus continuous with the stem (§ *Eumorchella*). The former section includes the two closely allied species *M. hybrida* and *M. rimosipes*. The latter section contains a number of species, some of which are apparently well marked and others appear to be growth modifications of the common *M. esculenta* which we have selected for a frontispiece. The American species deserve careful study and comparison in the field.

Gyromitra contains seven species some of which are the largest members of the order and perhaps of the entire class Ascomycetes.

G. esculenta is eaten with us, and in Germany it is canned under the name "Morcheln"; *G. brunnea* is a much larger species from Indiana, Ohio and Kentucky.

LITERATURE.

Schroeter. Die natürlichen Pflanzenfamilien, 1¹: 162–172.

Rehm. Rabenhorst's Kryptogamen flora Deutschlands, u. s. w. 1³: 1134–1208.

Saccardo. Sylloge Fungorum, 8: 7–53; 10: 1–3; 11: 391–393.

Cooke. Mycographia, 1–10, 87–104, 179–206, 215–220. *Pl.* 1–4, 41–46, 81–96, 101, 102. 1879.

Massee. A Monograph of the Geoglosseae. Ann. Bot. 11: 225–306. *Pl. 12, 13*. 1897.

Underwood. On the Distribution of the North American Helvellales. Minn. Bot. Studies, 1: 483–500. 1896.

Burt. A list of the Vermont Helvelleae, with descriptive notes Rhodora, 1: 59–67, *Pl. 4*. 1899.

The Lichens are ascomycetous fungi enclosing in their mycelium unicellular or filamentous algae on which they live parasitically. On account of a supposed symbiosis of the two groups in the same individual lichen, there has been a hesitancy to place the lichens in their true position in the system among the Pezizales, Sphaeriales and other orders of the Ascomycetes. There is every gradation from forms regularly enclosing algae to those in which algae are only incidentally present, and the strongest reasons for keeping lichens in a distinct class are custom and convenience, neither of which should have weight in a natural system.

A few basidiomycetous fungi also enclose algae in a similar way, so that there are *Ascolichenes*, *Hymenolichenes* and *Gastrolichenes*. The greater part of our familiar species are allied to the Pezizales.

CHAPTER VI

THE FUNGI IMPERFECTI

Besides the true Ascomycetes there are a large series of forms that are analogous in some cases to the known conidial stages of ascomycetes. Some of these are undoubtedly the conidial form or stage of some ascomycetous species, but the relation between conidial and ascosporic forms has not yet been discovered; others may at some time have been thus connected but having become perennial they have lost their ascomycetous form; others still are without doubt perfect species whose relations have not yet been determined. The species classed in this extensive group are arranged in three orders as follows:

1. Conidia in perithecia-like cavities (pycnidia).   1. **Sphaeropsidales.**
Conidia superficial, borne on loose or innate hyphae; no pycnidia. 2.
2. Hyphae innate with the matrix   2. **Melanconiales.**
Hyphae somewhat superficial, often floccose.   3. **Moniliales.**

Except in the case of certain genera of leaf parasites, and a few species injurious to cultivated plants, no considerable amount of study has been given to the American species of these orders. Our knowledge of them consists of miscellaneous descriptions of species whose characters have never been compared with each other. The added difficulty exists in these orders, that many of its members are only stages in the life history of some fungus existing under another name. This relation is difficult to discover and must be discovered often by a combination of culture experiments with a careful study of its occurrence in the field. While we know that many of these so-called genera are merely form-genera, we are obliged to treat them as we do other genera until the true relations of their last remaining species has been worked out. It would be a valuable aid to their study if compiled descriptions even of American genera and species were accessible to students. As it is, we are practically forced to rely on a Latin compilation not only of the American species but all the others from the known world. Nearly four hundred and fifty named genera are included in the three orders.

Order 1. SPHAEROPSIDALES.

Among the fungi imperfecti the Sphaeropsidales are so named because of the fact that they produce structures resembling the perithecia produced by the Sphaeriales. These perithecia like bodies are the pycnidia (*Pl. 5, f. 2*), and the spores instead of developing in asci are produced from the walls of the pycnidium direct. The group is a very extensive one including a number of leaf-spot diseases of cultivated plants. Among these are those of the apple, catalpa, maple, celery, sweet potato and violet caused by species of *Phyllosticta*; those of currants, raspberries, carnations, horseradish and lettuce caused by species of *Septoria*; that of the rose caused by *Actinonema*, and a number of others. While the greater number of the species are saprophytic on stems, branches, etc., there are extensive genera of parasitic forms. Among these are *Phyllosticta* with over 400 species and *Septoria* with over 600 species, a fair proportion of which are found in America. In most of the species the pycnidia are black, but in the small family Zythiaceae they are light colored.

The four families can be distinguished as follows:

1. Pycnidia globose, conic or lenticular.   2.
   Pycnidia more or less dimidiate, irregular or shield-shaped, black.
                                                   Leptostromataceae.
   Pycnidia cup-shaped or patelliform, black.   **Excipulaceae.**
2. Membranous, carbonaceous or coriaceous, black. **Sphaeropsidaceae.**
   Fleshy or waxy, light colored.   **Zythiaceae.**

Among the SPHAEROPSIDACEAE the larger genera with simple hyaline spores are *Phoma* and *Vermicularia* which are saprophytic, and *Phyllosticta* parasitic on leaves. *Ampelomyces* is a minute form parasitic on the hyphae of various Erysibaceae.[*] (*Pl. 5. f. 3.*) *Sphaeropsis* has simple brown spores and *Diplodia* has two-celled brown spores.

Among the genera with hyaline two-celled spores are *Ascochyta* and *Actinonema*, one of the species of the latter forming black

---

[*] *Cf.* Griffiths. The common Parasite of the Powdery Mildews. Bull. Torrey Bot. Club, **26**: 184-188. *pl. 358.* 1899.

radiating spots on the leaves of roses. *Darluca filum* with similar spores is a common parasite on the rusts (Uredinales). *Septoria*, one of the largest genera in the family, has slender filiform or rod-like spores which are either divided into numerous cells by cross septa or are marked with clear dots (guttulate) (*Pl. 5. f. 1, 2*).

Among the eighty-seven genera of this family the following leaf parasites can be distinguished by this brief synopsis:

1. Spores 1-celled, ovoid or oblong, hyaline. PHYLLOSTICTA.
   Spores 1-celled, ovoid or oblong, fuscous or smoky. 2.
   Spores 2-celled, hyaline. 4.
   Spores with transvers septa only, 3-many-celled, oblong or fusiform. 5.
   Spores muriform, ovoid or oblong; pycnidia subcutaneous or erumpent. CAMAROSPORUM.
   Spores elongate-fusiform, continuous or septate. 6.
2. Pycnidia smooth. 3.
   Pycnidia setose. CHAETOMELLA.
3. Spores large (15–30µ long) stipitate. SPHAEROPSIS.
   Spores smaller (3–10µ long) scarcely stipitate. CONIOTHYRIUM.
4. Pycnidia on definite discolored spots. ASCOCHYTA.
   Pycnidia not on definite spots, with a radiate arachnoid subiculum. ACTINONEMA.
5. Spores smoky or olivaceous, muticous; pycnidia subcutaneous. HENDERSONIA.
   Spores smoky or olivaceous, ciliate. CRYPTOSTICTIS.
   Spores hyaline; pycnidia subglobose. STAGNOSPORA.
6. Pycnidia usually on definite spots; spores usually very narrow. SEPTORIA.
   Pycnidia rarely on spots; spores thickened. PHLEOSPORA.

The remaining genera are mainly saprophytic, growing in similar situations as the Sphaeriales and like them existing free, imbedded and erumpent from beneath the bark of stems or twigs, or they may even be imbedded in a stroma. Since the ascus in the Sphaeriales is sometimes a somewhat evanescent structure, it is occasionally difficult in practice to distinguish members of this family from the Sphaeriales. The main structural difference between pycnidia and perithecia is found in the method of bearing the spores directly from the walls in the former and in distinct asci in the latter; usually by suitable choice of material, however, it will be possible to demonstrate the presence of asci in those species in which they are normally produced so that one need not ordinarily be in doubt as to the true position of the plant under examination.

The striking similarity of certain of the Sphaeropsidaceae to the Sphaeriales leads one to suspect that in some cases the former are really members of the Sphaeriales in which the ascus has become abortive. In a number of species, however, the true ascosporic condition is known in addition to the pycnidial stage, and probably a similar relation exists between many species of Sphaeropsidaceae on the one hand, with described Sphaeriales on the other. The discovery of such relations is no simple matter, and with our present knowledge of methods of cultivation, the certain demonstration of relationship is often impossible.

The ZYTHIACEAE * with light colored pycnidia bear the same relation to the Hypocreales, that the last family does to the Sphaeriales. The species of the fourteen genera are mainly saprophytic. The genera are mostly small, *Zythia*, the largest having ten species, only two of which are reported from this country with a third from Cuba.

The LEPTOSTROMATACEAE with shield-shaped pycnidia, while not distinctively leaf-parasites, include numerous species among the thirteen genera that inhabit the living or languid leaves of some of the higher plants The genera containing such leaf-parasites may be distinguished as follows :

1. Spores one-celled, hyaline, globose or ellipsoid. 2.
   Spores 2-many-celled, hyaline, fusiform. DISCOSIA.
   Spores 4-celled, the cells arranged in the form of an irregular cross.
   ENTOMOSPORIUM.
2. Walls of the pycnidia distinctly parenchymatous. 3.
   Walls of the pycnidia not parenchymatous. SACIDIUM.
3. Pycnidia splitting longitudinally, somewhat hysterioid. 4.
   Pycnidia without a mouth, never splitting longitudinally. 5.
4. Pycnidia lanceolate or elongate. LEPTOSTROMA.
   Pycnidia nearly circular. LABRELLA.

---

* This family has commonly been known as Nectrioideae, but as this name is not derived from that of a representative genus of the family, and further does not have the regular family ending, we make the above substitution.

5. Pycnidia shield-shaped, separating readily; basidia obsolete.
LEPTOTHYRIUM.
Pycnidia irregular; basidia columnar. PIGGOTIA.
Pycnidia irregular; basidia obsolete. MELASMIA.

Entomosporium (*Pl. 5, f. 4*), so-called from the resemblance of the ciliated spores to insects, is a common parasite of pears and quinces; a species of *Piggotia* is more or less abundant on the ash; and a species of *Melasmia* is found on the American elm.

The EXCIPULACEAE are cup-shaped or lenticular, either membranous or carbonaceous, smooth or pilose; in exceptional genera the pycnidia are more or less elongate, simulating species of the Hysteriaceae. Twenty-two genera are known, most of which are small, except *Excipula* and *Dinemosporium* with one-celled hyaline spores, the former with smoothish pycnidia and the latter with pilose pycnidia and spores ending in a bristle. (*Pl. 5. f. 5.*) Both the genera named are among the few containing leaf-parasites.

LITERATURE.

Saccardo. Sylloge Fungorum, 3: 3-695; 10: 100-444; 11: 472-561.

Martin. The Phyllostictas of North America. Jour. Mycol. 2: 13-20, 25-27. 1886.

———— Enumeration and Descriptions of the Septorias of North America. Jour. Mycol. 3: 37-41, 49-53, 61-69, 73-82, 85-94. 1887.

Allescher. Rabenhorst's Kryptogamen Flora Deutschlands, u. s. w. 1^6: 1– . 1898. (Current.)

The parts of Die natürlichen Pflanzenfamilien relating to the Fungi Imperfecti have not yet appeared.

Order 2. MELANCONIALES.

The order Melanconiales contains a single family, the MELANCONIACEAE, made up of a comparatively small number of species in which neither asci nor pycnidia are developed. As a rule the spores are borne in cavities without special walls, often rising from little masses of short hyphae. The greater number of the species are saprophytic on decaying vegetable substances but a few are parasitic and occasion a number of destructive diseases.

The diseases caused by the fungi of this order are known as *anthracnose* and are chiefly produced by species of the genera *Glocosporium* and *Colletotrichum*. Among these are the anthracnose of the bean, cotton, hollyhock, melon, pepper, spinach, tomato and watermelon caused by members of the latter genus; and those of the apple,[*] blackberry, cucumber, currant, peach, a second on pepper, persimmon, raspberry, rose and a second on tomato caused by species of *Glocosporium*. The anthracnose of the chestnut is caused by a species of *Marsonia*.

Of the thirty six genera the following are the principal ones containing leaf parasites:

1. Conidia 1-celled, globose, ovoid or short cylindric, hyaline. 2.
    Conidia filiform, often twisted; acervuli pale. CYLINDROSPORIUM.
    Conidia fusiform-falcate; acervuli black or gray. CRYPTOSPORIUM.
    Conidia 2-celled, solitary. MARSONIA.
    Conidia 3-many-celled, smoky. 5.
    Conidia 3-many-celled, hyaline; acervuli pale. SEPTOGLOEUM.
2. Conidia with branched bristles at apex. PESTALOZZIELLA.
    Conidia not spiny. 3.
3. Acervuli bristly at the margins. COLLETOTRICHUM.
    Acervuli not provided with bristles. 4.
4. Acervuli soon erumpent, minute, bright colored. HAINESIA.
    Acervuli long covered, gray or pallid, somewhat waxy. GLOEOSPORIUM.
5. Conidia with hyaline cilia at apex. PESTALOZZIA.
    Conidia oblong, not ciliate or rostrate. CORYNEUM.

Of the above genera *Glocosporium* is the largest with some one hundred and twenty five species and *Pestalozzia* follows next with about ninety. Besides the leaf parasites a common twig parasite may be seen on young twigs of *Cornus* causing them to turn yellow and die, when they are covered with the erumpent acervuli of *Myxosporium nitidum*.

LITERATURE.

Saccardo. Sylloge Fungorum, 3: 696–812; 10: 446–509; 11: 562–585.

Ellis & Everhart. The North American Species of Glocosporium. Jour. Mycol. 1: 109–120. 1885.

---

[*] Also called ripe rot.

———— North American Species of Cylindrosporium. Jour. Mycol. 1: 126-128. 1885.

———— Additions to Cercospora, Gloeosporium and Cylindrosporium. Jour. Mycol. 3: 13-22. 1887.

Stoneman. A comparative Study of the Development of some Anthracnoses. Bot. Gaz **26**: 69-120, *pl. 7-18.* 1898.

Order 3. MONILIALES.

The largest order of the fungi imperfecti is the group of fungi often known as the Hyphomycetes, and called "Hyphos" for short. In this group a large number of forms have been placed that have later been found to represent merely the conidial stage of ascomycetous fungi. Among these is the genus *Oidium* which contains mildews that have developed only a conidial reproduction. As some of the members of these various form-genera are not yet identified with the ascomycetous condition, it is necessary to retain them as a matter of convenience, although we know that they cannot rank as genera in any strict biological sense. There are also numerous forms that are only partially known and these have been placed in this group temporarily as a matter of convenience. In fact the order has been a convenient catch-all for nondescript fungi of all sorts, so that the order has become a sort of by-word among mycologists. But besides these really imperfect forms there are a large number that probably represent forms that are unconnected with any other stage of growth and are thus in themselves perfect fungi; they are simply in bad company. By far the greater part of the order are saprophytic but a considerable number are parasitic; among the latter is the genus *Cercospora* in which over 450 species have been described, a large part from the United States. Here also belongs the smut of *Sporobolus Indicus* which has taken the name of smut grass in the South from the almost universal prevalence of the fungus.*

A few fungi injurious to cultivated plants are found to be due to species of this order and among them some of the most virulent of their kind. Among these are the various leaf-blights due to species of *Cercospora*, *Ramularia*, *Macrosporium* and *Helmintho-*

---

* This smut must not be confused with the ordinary smuts of grain and grasses which are members of the order Ustilaginales.

sporium which affect a large number of cultivated plants. Here also are to be found the causes of various diseases known as scab. Among these are the black blotches on apples and pears due to species of *Fusicladium;* the scab of potatoes and beets caused by *Oospora scabies;* the scab of the fig caused by a *Fusarium;* and the leaf blights of the plum, peach, cherry, tomato and spinach caused by species of *Cladosporium.* To the same order also belongs the very destructive rot of plums and peaches which probably causes as much loss to fruit growers as all other diseases of these fruits combined. This is caused by *Monilia fructigena* (*Pl. 5. f. 9*).

The order is made up of four families which may be distinguished as follows:

1. Hyphae more or less floccose or mould-like.   2.
   Hyphae closely united into an elongate, columnar fascicle. **Stilbaceae.**
   Hyphae closely conglutinate in a tubercular mass. **Tuberculariaceae.**
2. Pale or light colored, often quickly collapsing.   **Moniliaceae.**
   Fuscous or black, usually rather rigid.   **Dematiaceae.**

Among the many genera of the MONILIACEAE,* *Ramularia* with ovate-cylindric conidia is probably the largest genus, in which over one hundred species have been described; they are mainly parasitic on leaves (*Pl. 5. f. 11*). The genera containing species parasitic on leaves may be distinguished as follows:

1. Spores varying from spherical to short cylindric, one-celled, hyaline. 2.
   Spores 2-celled, hyaline or light colored.   5.
   Spores 3–many-celled, hyaline or light colored.   6.
2. Hyphae very short, its cells scarcely distinct from the conidia.   3.
   Hyphae elongate, distinct from the smooth conidia.   4.
3. Conidia borne in heads.   (Parasite on *Cornus Canadensis.*)
                                                              GLOMERULARIA.
   Conidia in chains, lemon shaped, very large; hyphae branching (often
      saprophytic).   MONILIA.
   Conidia in chains, ellipsoid, flattened at base.   OIDIUM.
4. Conidia globose or ovoid, on ascending fertile branches (mostly sapro-
      phytic).   BOTRYTIS.
   Conidia on suberect branches which are denticulate above. OVULARIA.

---

*This family called by Saccardo the Mucedineae from their resemblance to the true phycomycetous moulds (*Mucor mucedo, et. al.*) is given the above name in accordance with the principle stated on p. 19, and as followed in the Zythiaceae (note, p. 71).

5. Hyphae spirally twisted. BOSTRICHONEMA.
   Hyphae not spirally twisted. DIDYMARIA.
6. Conidia ovate-cylindric. RAMULARIA.
   Conidia obclavate-pyriform. PIRICULARIA.
   Conidia vermiform or filiform. CERCOSPORELLA.

Of the above genera, *Botrytis* is often seen growing on dying plants in greenhouses, particularly on slips just planted which have been kept overwarm and moist; it also appears to be somewhat parasitic (*Pl. 5. f. 10*). Most of the genera are small. *Oidium* is made up mostly of conidial stages of the Erysibaceae. Besides the genera mentioned there are some eighty-five others mostly saprophytic, almost any of which may be found with us, though many have not yet been reported as American.

The family DEMATIACEAE with fuscous or black floccose hyphae is a large one; while the spores are occasionally nearly or quite hyaline the hyphae are always fuscous or dark brown. The genera containing leaf parasites can be distinguished by the following synopsis:

1. Conidia 1 celled, ovoid or oblong, blackish or sub-hyaline on fuscous hyphae. 2.
   Conidia 2-celled, ovoid or oblong. 6.
   Conidia varying from ovoid to vermiform, 3-many-celled by septa at right angles to the long axis, fuscous. 8.
   Conidia globose or oblong, muriform (dictyoid). 13.
2. Hyphae short, scarcely distinct from the globose or ovoid conidia. 3.
   Hyphae longer, distinct from the conidia. 4.
3. Conidia solitary. CONIOSPORIUM.
   Conidia in chains. TORULA.
4. Conidia fuscous. 5.
   Conidia hyaline on fuscous hyphae, erumpent, fusoid. ELLISIELLA.
5. Hyphae creeping, with curved branches. CAMPSOTRICHUM.
   Hyphae erect, fasciculate, rather short. HADROTRICHUM.
6. Hyphae erect, short, somewhat fasciculate. 7.
   Hyphae erect, short, flexuose (parasitic on clover). POLYTHRINCIUM.
   Hyphae more or less spreading, branched; conidia often in short chains (more often saprophytic). CLADOSPORIUM.
7. Conidia borne only at apex. FUSICLADIUM.
   Conidia both apical and lateral. SCOLECOTRICHUM.

8. Fertile hyphae short or only slightly distinct from the solitary conidia. 9.
   Hyphae longer and distinct from conidia. 11.
9. Conidia smooth. 10.
   Conidia with a slender appendage. CERATOPHORUM.*
10. Conidia cylindric (more often saprophytic). CLASTEROSPORIUM.
    Conidia ovoid, in tufts. STIGMINA.
11. Conidia smooth, elongate or vermiform. 12.
    Conidia echinulate, oblong. HETEROSPORIUM.
12. Hyphae rigid (more commonly saprophytic). HELMINTHOSPORIUM.
    Hyphae soft, simple or branched, often forming leaf-spots.
    CERCOSPORA.
13. Conidia solitary; hyphae erect, somewhat fasciculate, soft.
    MACROSPORIUM.
    Conidia in chains; hyphae velvety, erect. ALTERNARIA.
    Conidia in chains; hyphae crustaceous or interwoven. FUMAGO.

Of the above genera some are very numerous in nominal species but it is more than probable that many of the described species will prove identical after they have had more careful study and cultivation. *Cercospora* is the largest genus of strictly leaf parasites. Of over four hundred and fifty species, nearly half are found in the United States, nearly every family of flowering plants furnishing one or more hosts. The genus is specially abundant in the Southern States (*Pl. 5. f. 12*). Besides *Cercospora*, *Macrosporium* has over eighty species, *Cladosporium* has over a hundred, and *Helminthosporium* over a hundred and twenty-five. All these three genera, however, are more commonly saprophytic and usually appear as leaf fungi only after the death of the tissues. The species of *Fumago* are doubtless related to *Capnodium*, as the ascosporic stage.

Besides the genera which are regularly or occasionally parasitic on leaves there are over one hundred others which are normally saprophytic on dead wood and other substrata. Among the forms with simple spores, *Streptothrix* with short twisted hyphae often forms numbers of brown rounded heaps on the bark of fallen linden trees and sometimes those of other species. *Glenospora Curtisii* is also common in the South, forming black patches on living twigs and branches of *Magnolia* and other trees.

*Cf. Pl. 5. f. 14.

The STILBACEAE are mainly saprophytic. Among the genera with simple hyaline conidia *Isaria* contains several species parasitic on insects which probably represent the conidial stage of species of *Cordyceps*. The family contains about sixteen more or less clearly defined genera.

The TUBERCULARIACEAE, with over forty genera, are also chiefly saprophytic in habit. *Fusarium* is the largest genus with over one hundred and eighty species. Some of our worst diseases of cultivated plants are due to species of this genus, particularly in the Southern States ; among these are the wilt of cotton, watermelon and cow peas. *Fusarium miniatum* forms the red slimy fungus that often appears on the sap oozing in spring from stumps of recently cut trees. It often grows in masses on the trunks of the ironwood (*Ostrya*) whose bark has been pierced by woodpeckers.

The species of *Tubercularia* from the conidial stages of species of *Nectria* among the Hypocreales. Some of the species of *Illosporium* form pink parasites on various foliaceous lichens. *Tuberculina persicina* is a common parasite on various Uredinales causing the sori to assume a purplish tint.

The entire order of Moniliales thus forms an immense heterogeneous and imperfectly known group of fungi which has long been the *bête noire* of mycologists. As will be seen from the citations of American literature only a few of the more conspicuous genera of leaf-parasites have ever been studied comparatively and the field is open for much serious study.

LITERATURE.

Saccardo. Sylloge Fungorum, 4: 1-807; 10: 510-739; 11: 586-656.

Corda. Icones Fungorum, 1-6 : *64 pl.* 1837-1854.

Ellis & Everhart. Enumeration of the North American Cercosporae. Jour. Mycol. 1 : 17-24, 33-40, 49-56, 61-67. 1885.

——— North American species of Ramularia. Jour. Mycol. 1 : 73-83. 1885.

——— Supplementary enumeration of the Cercosporae. Jour. Mycol. 2 : 1-2. 1886.

——— Additions to Cercospora, Gloeosporium and Cylindrosporium. Jour. Mycol. 3 : 13-22. 1887.

——————— Additions to Ramularia and Cercospora. Jour. Mycol. 4 : 1–7. 1888.

Atkinson. Some Cercosporae from Alabama. Journ. Elisha Mitchell Sci. Soc. 8 :—(1–35). 1892.

Morgan. North American Helicosporae. Journ. Cincinnati Soc. Nat. Hist. 15: 39–52.

——————— Two new genera of Hyphomycetes. Bot. Gaz. 17 : 190–192. 1892.

Thaxter. On certain new or peculiar North American Hyphomycetes. I. Bot. Gaz. 16 : 15–26 1891.

Pettit. Studies in artificial Cultures of Entomogenous Fungi. Bull. Cornell Univ. Agric. Exper. Sta. 97 : 339–378. *IV. 1–11.* 1895.

Pound and Clements. A Rearrangement of the North American Hyphomycetes. Minn. Bot. Studies, 1 : 644–673. 1896; 726–738. 1897.

The above papers are nearly all descriptive, giving more than mere enumerations of species. No general work on the American species of the order has yet appeared, and such a work is greatly needed.

CHAPTER VII

THE LOWER BASIDIOMYCETES

(*Rusts and Smuts*)

The third class of fungi known as the Basidiomycetes contains two series of organisms very dissimilar in habit, the first series parasitic on plants, and the second series forming the saprophytic forms of mushrooms and puff balls that constitute not only the highest types of fungi but at the same time those that are popularly most widely known. The class Basidiomycetes is composed of twelve orders which may be separated as follows:

1. Parasitic on spermaphytes, often deforming the host; mostly inconspicuous; (rusts and smuts). 2.
   Saprophytic, mostly gelatinous, fleshy or woody fungi, usually conspicuous; terrestrial or epixylous, rarely parasitic on other fungi of their own class. 3.
2. Producing in ovaries or leaves, smut-like chlamydospores from which the basidia-like conidiophores arise in germination. 1. **Ustilaginales.**
   Producing black, brown or yellow rust-like pustules under the epidermis of leaves or stems, or developing clusters of crater-like openings with spores formed in chains inside a membranous pseudoperidium.
   2. **Uredinales.**
   Producing whitish spots, pustules or inflated galls mostly on Ericaceae; basidia clavate with small sterigmata. 6. **Exobasidiales.**
3. Gelatinous fungi with divided basidia. 4.
   Gelatinous fungi with long clavate two-forked basidia; spores dividing before germination. 5. **Dacryomycetales.**
   Fleshy, coriaceous or woody fungi with undivided basidia. 5.
4. Basidia divided crosswise (septate). 3. **Auriculariales.**
   Basidia divided obliquely or lengthwise, commonly into four parts.
   4. **Tremellales.**
5. Spores arising from basidia which form a distinct membranous hymenium, naked at maturity, and covering the surface of gills, pores, or spines. (*Hymenomycetes;* mushrooms, bracket fungi.) 7. **Agaricales.**
   Spores arising from basidia enclosed within a definite peridium. (*Gasteromycetes.*) 6.

6. Spores borne in a more or less deliquescent gleba which is at first enclosed in an egg-like sac (peridium), but at maturity elevated on an elastically expanding receptacle; (stink horns).   8. **Phallales.**
Spores remaining within the peridium until maturity   7.
7. Basidia united into a hymenium which lines the walls of irregular cavities.   8.
Basidia uniformly distributed through the peridium, or forming skein-like masses.   12. **Sclerodermatales.**
8. Hymenial cavities remaining together within the peridium, their boundaries mostly disappearing at maturity.   9.
Hymenial cavities (sporangioles) separating at maturity from the cup-like peridium; (bird's nest fungi).   11. **Nidulariales.**
9. Remaining fleshy until the maturity of the spores; no capillitium.
    9. **Hymenogastrales.**
Fleshy when young, at maturity filled with dust-like spore masses mixed with capillitium; (puff-balls).   10. **Lycoperdales.**

The above grouping is necessarily artificial and should be contrasted with the following more natural arrangement:

1. *Hemibasidii:*   Ustilaginales.
2. *Protobasidii:*   Uredinales, Auriculariales, Tremellales.
3. *Autobasidii:*   All the remaining orders.

The relations of some of the lowest types are not yet clearly known, and the relations of the entire group to other plants are even more obscure.

Order 1. USTILAGINALES.

The smuts form a group of parasitic fungi too well known to the farmer since they result in the loss of a large percent of his crops annually, notably among the cereals. The common corn smut is a familiar example. It commences to make its appearance as small distorted nodules either in the young kernels, in the tassels, or quite frequently at the joints of the stems. These nodules increase in size, becoming spongy, white, and glistening; spore formation is progressing, meanwhile, and the entire tissue soon becomes filled with a black mass of greasy spores which form the reproductive bodies of the fungus. Other species are parasitic in the heads of wheat, barley and oats; others still are found in the leaves of various grasses forming long lines of black spores and externally resembling black rust, but readily distinguished by the form of the spores ; still others are found in the ovaries of various plants, and occasionally in other parts. A few have light colored spores, among them one affecting the ovaries of *Oxalis*; another species is parasitic in the anthers of pinks. The spores thus formed, commonly called chlamydospores, germinate, developing a short promycelium and forming the spores proper which in turn produce the mycelium of the fungus. In the case of the common smuts of cereals the smut spores lie dormant in the furrow of the kernel until the germination of the seeds ; the fungus then germinates producing its spores, whose mycelium penetrates the young plantlet and extends throughout its tissues. As a rule, a parasite of this kind shows little evidence of its presence in the host plant until the spores commence to form, usually in the young ovaries where large quantities of available nutriment are being carried for the supreme effort of flowering and producing seed. This nutriment the fungus appropriates and there develops its spores. Two families are distinguished by the method of bearing spores from the germination of the chlamydospores as follows :

Chlamydospores germinating with a several-celled promycelium which bears the spores at the ends of the cells. (*Pl. 6. f. 1, 2.*)
Family 1. **Ustilaginaceae.**

Chlamydospores germinating with an undivided promycelium which bears the spores in a cluster at the apex. (*Pl. 6. f. 3.*)
Family 2. **Tilletiaceae.**

Family 1. **Ustilaginaceae.**

The smuts of this family except those of certain genera cannot be distinguished by their gross appearance or by their microscopic characters from the members of the next family. The biological characters of germination and spore production must be resorted to in order to distinguish them. Some germinate soon after production, some germinate after a period of rest, and some retain their power of germination for several years but this is somewhat unusual. Their cultivation is a simple matter when conducted in a drop culture in nutrient fluid. The genera are distinguished as follows :

1. Chlamydospores simple. 2.
    Chlamydospores united in 2's by a narrow isthmus. SCHIZONELLA.
    Chlamydospores in masses of three or more. 3.

2. Promycelium 1-5-celled, bearing lateral and terminal spores; (parasitic on various hosts). USTILAGO. *

Promycelium 2-celled, each bearing a number of spores on steriginata; (parasitic on Cyperaceae). ANTHRACOIDEA.

3. Spore-masses composed of a few cells; (a single Californian species on *Atriplex*.) POIKILOSPORIUM.

Spore-masses composed of numerous clustered cells. 4.

4. Chlamydospores loosely united, often falling apart at maturity. SOROSPORIUM. †

Chlamydospores closely united together. 5.

5. Promycelium with both lateral and terminal spores. TOLYPOSPORIUM.
Promycelium branched, with single lateral spores. TOLYPOSPORELLA.
Promycelium with a single terminal spore. THECAPHORA.

Of the above genera, *Ustilago* is by far the largest, containing nearly two hundred species of which at least one-third are American. (*Pl. 6. f. 1, 2*.) *U. zeae* forms the smut of corn, *U. avenae* the common smut of oats, *U. tritici* the common smut of wheat; *U. longissima* forms long brown lines on the leaves of *Glyceria;* *U. violacea* appears in the anthers of *Dianthus* and other members of the pink family, *U. oxalidis* in the ovaries of *Oxalis*, and many other grasses and dicotyledonous plants are similarly affected. Our American species need a careful comparative study, a statement equally true for the entire order.

Schizonella is represented by a single species parasitic on the leaves of *Carex Pennsylvanica* forming narrow interrupted black lines in early spring. *Sorosporium* has several species on grasses, besides *S. saponariae* found in the ovaries of *Saponaria* and other pinks. *Tolyposporella* has a single species formed underneath the leaf sheath of *Chrysopogon nutans*, *Tolyposporium* has several species on *Juncus* and various grasses, while *Thecaphora* has a few species on *Convolvulus* and various members of the pea family.

Family 2. **Tilletiaceae.**

This family contains genera still more diverse from each other. While some like *Tilletia* and *Urocystis* are similar in general ap-

---

*The genus *Cintractia* is sometimes separated from *Ustilago*.

† In this condition there is often difficulty in distinguishing this genus from *Ustilago*.

pearance to members of the preceding family, *Entyloma* and other genera bear little superficial resemblance to ordinary smuts and for them this name seems a misnomer. The genera may be distinguished by the following synopsis :

1. Chlamydospores simple ; spores in a terminal cluster of not more than twelve.   2.
   Chlamydospores compound, several or many cells united in a mass.   3.
2. Chlamydospores in dusty masses mostly dark-colored.   TILLETIA.
   Chlamydospores formed in small or medium sized often whitish spots sunken into the tissues of the host, mostly light colored.   ENTYLOMA.*
3. All the cells of the mass similar and fertile ; (parasitic on *Trientalis*).
     TUBURCINIA.
   Outer cells sterile, or the outer fertile with sterile cells within the mass.   4.
4. Spore cells few; the outer cells sterile.   UROCYSTIS.
   Spore cells numerous ; the outer cells sterile.   DOASSANSIA.
   Spore cells numerous ; a single or double layer of outer cells fertile.   5.
5. Interior of spore-mass filled with closely packed parenchymatous cells.
     DOASSANSIOPSIS.
   Interior of spore-mass hollow containing a network of united hyphae ; (parasitic on *Lemna*).   CORNUELLA.

Of these genera, *Tilletia* contains a considerable number of species on various grasses and cereals. (*Pl. 6. f. 3.*) Among these are *T. tritici* forming the bunt or stinking smut in the heads of wheat, *T. secalis* found on rye, and *T. striaeformis* forming parallel brown lines on the leaves of various grasses.†

Entyloma is represented by numerous species on *Menispermum*, *Physalis*, *Myosotis*, *Ranunculus* and various other plants.

Urocystis is a typical smut and has a dozen or more American species. (*Pl. 6. f. 4.*) *U. agropyri* forms parallel lines on the leaves of quack grass, *U. occulta* appears in the stems and leaves of rye, *U. anemones* is common in the leaves of *Hepatica*, *U. cepulae* forms the destructive onion smut and there are others.

---

* Farlow (Bot. Gaz. **8** : 271-278. 1883) gives a synopsis of the United States of *Entyloma* then known ; several others have since been described.

† We have in this country an *Ustilago*, a *Urocystis* and a *Tilletia*, all possessing this habit. The fungus in each case apparently follows the trend of the parallel venation of the host.

Doassansia has several species mostly in marsh or water plants like *Alisma* and *Sagittaria*, the latter supporting two species. *Doassansiopsis* affects similar plants; among them *Sagittaria* has two species, one forming gall-like bodies on the stems; other species are found in *Potamogeton* and *Glyceria*.

The entire order forms a most interesting economic group. With the exception of *Doassansia* and its allies, the American species have never been monographed. The biological study of certain species has been very extensive.

LITERATURE.

Saccardo. Sylloge Fungorum, 7: 449-527; 9: 282-291; 11: 174-230.

Winter. Rabenhorst's Kryptogamen Flora Deutschlands, u. s. w. 1¹: 80-131.

Plowright. A Monograph of the British Uredineae and Ustilagineae, 58-118, 272-301. 1889.

Brefeld. Untersuchungen aus dem Gesammtgebiete der Mykologie, 11 and 12: 1895.

Setchell. Preliminary Notes on the Species of *Doassansia* Cornu. Proc. Am. Acad. Sci. 26: 13-19. 1891.

—— An Examination of the Species of the Genus *Doassansia* Cornu. Ann. Bot. 6: 1-48. *Pl. 1, 2.* 1892.

Dietel. Die natürlichen Pflanzenfamilien, 1¹**: 2-24. 1897.

Order 2. UREDINALES.

The order Uredinales contains the largest array of parasitic fungi of any existing order and one in which there is an extensive economic interest because of the fact that several very important products are diminished by the ravages of these fungi; wheat and oats among the cereals, apples and quinces among the fruits, roses and carnations among the ornamental plants, are only a few examples of plants of economic importance that are parasitized by members of this order.

The rusts are further interesting because of the peculiar change of host which takes place in certain of the species. To illustrate this habit, the somewhat familiar examples of wheat-rust and orchard-rust may be cited since the two parasites come to be of economic importance in different stages of their life history, simply from the fact that the host plants affected happen to be of importance themselves and the fungus in either stage detracts from its free growth. The common wheat-rust commences its life history as a parasite on the barberry (*Berberis*); in May or June thickened yellow spots appear on the leaves of the barberry and from these a series of rounded yellowish bodies push their way through the epidermis and open up as little cups or craters; from their habit of growth they have taken the name of cluster-cups. Before their true relations were known they were described as definite fungus species under the name of *Aecidium*. The spores are produced in chains, numbers of which are packed so closely within the membranous pseudoperidium which covers them, as to render them angular from pressure. These spores are one-celled and thin-walled. They germinate in the presence of moisture, but will come to naught unless they are carried by the wind or otherwise to young plants of the wheat which they parasitize, the mycelium from the germinating spore entering the host plant through one of the stomata. Once inside, growth takes place and in due time the mycelium accumulates under the epidermis and there produces a mass of spores whose growth finally ruptures the epidermis and appears as a *sorus* of one-celled, yellowish-brown or rusty spores which have a deciduous stalk, and are made up of a single cell. (*Pl. 6. f. 5.*) This stage is known as red rust by the farmers and, like the cluster cup, was described, before its relations had been made out, as a member of a fungus genus, *Uredo*. It is by means of these thin-celled summer spores that the wheat-rust makes such havoc in a wheat field. The spores are loosened from their stalks at maturity and are carried by the wind to other wheat plants; a few favorable sultry days will furnish the conditions necessary for rapid germination and what was a single centre of infection has become a thousand, each rapidly producing new crops of spores and continuing ever to widen the amount of infection; should this attack come at the time when the young kernels of wheat were forming, the nutrition that would naturally go to them would be absorbed by the mycelium of the fungus and the kernels would become shriveled and worthless. A little later either from the same sori in which the red rust spores were produced or in others, black spores appear. These differ from the preceding (1) In having a permanent stalk; (2) In having two cells instead of one, and (3) In having thick walls. (*Pl. 6. f. 6.*) These spores are known as the teleutospores and their function is normally that of a resting spore which will carry the life of the fungus over an unfavorable period, namely, the winter season. This black rust remains on the straw or stubble until the early warm days of the following spring, when the teleutospores germinate, developing a short promycelium on which as on basidia the fourth form of spores is produced. (*Pl. 6. f. 9.*) These so-called sporids are carried by the wind and if any of them chance to fall on the young leaves of the barberry just unfolding, they germinate, and their mycelium enters a stoma and becomes a parasite in the barberry which will in like manner develop cluster-cups as before. Thus in an endless cycle does the wheat rust carry on its round of miserable dependent existence. Other aecidial forms have a similar relation to species of the genus *Uromyces* which differs from *Puccinia* in having one-celled teleutospores. (*Pl. 6. f. 10.*)

Another case is seen in the orchard-rust. In summer many apple trees show bright yellow spots often a quarter of an inch wide on the surface of the leaves. This so-called rust is often so abundant that the foliage of the apple-tree has a distinct yellow tint in midsummer. A little later in the season these yellow spots produce a series of tubes very like the shorter ones of the cluster-cups except that they open by a series of chinks or fissures. Their spores scattered by the wind will germinate on the red cedar (*Juniperus*) and produce gall-like swellings, popularly known as cedar-apples which show themselves in early spring and later become covered with yellowish brown projecting masses of spores. As the later spring rains come on, these spore-masses swell into long gelatinous bodies in which the spores they contain germinate and produce the secondary spores in a manner quite analogous to the teleutospores of the wheat-rust. These sporids become free and may be carried back by the wind to the apple and there produce anew the centers of infection of the apple-rust. The more common cedar-apple, *Gymnosporangium macropus*, is an annual gall so that there is a necessity for a double transfer of the spores from host to host each season. In *G. globosum* the gall is perennial and successive crops of teleutospores are produced year after year in the same gall.

Not all rusts show this heteroecism, but produce their succes-

sive stages on the same host. We have a familiar instance in
the Northern States in the common mandrake or may-apple (*Podophyllum peltatum*); in the spring brilliant yellow cluster cups appear
on the young leaves in certain definite centres of infection; from
these the spores are scattered to other mandrake plants and in
early summer their bright green leaves become mottled with yellowish or brownish areas in which the teleutospores of the fungus
soon appear. (*Pl. 6. f. 8.*) In this rust there are no summer spores
(*Uredo*) since their function is carried on by the aecidiospores. In
other rusts as in *Puccinia graminella* teleutospores and aecidiospores are not only produced on the same plant, but from the
same mycelium. (*Pl. 6. f. 15.*)

Not all rusts possess aecidia. In some the teleutospores rarely
form owing to the fact that the mycelium has become perennial
probably in underground parts—a departure from the usual
habit of the fungus in which the mycelium is commonly confined
to a limited area not far from the point where the sorus is formed.
With the aecidia or sometimes independent of them a series of
structures are formed whose function is not known. These are the
spermogonia which appear to the naked eye as blackish points
usually on the opposite side of the leaf from the cluster cups. In
section they show pycnidia-like cavities from the mouths of which
small tufts of hairs emerge; within, the spore-like bodies (*spermatia*) are developed at the ends of slender mycelial threads.

The various habits of production of one, two or three forms of
reproductive bodies has led to the establishment of subgeneric
groups in some of the larger genera. For instance, in the genus
Puccinia the following groups have been proposed : *

EUPUCCINIA : producing aecidia, uredospores and teleutospores.

---

* Since in our own country alone there are some one hundred and
twenty-five aecidial forms whose relations to our 300 species of *Puccinia*
and 100 species of *Uromyces* are unknown, the position of any given species in these groups is, to say the least, very uncertain. There is abundant
opportunity for botanical students everywhere to institute a careful study of
the relations of the aecidia to teleutosporic forms. The study must be
taken up first in the field to establish suspicions from proximity of growth
and then to supplement field work with actual culture (inoculation) experiments.

PUCCINIOPSIS : producing aecidia and teleutospores.
BRACHYPUCCINIA : producing spermogones, uredospores and teleutospores.
HEMIPUCCINIA : producing uredospores and teleutospores.
MICROPUCCINIA : producing only teleutospores which remain over winter before germinating.
LEPTOPUCCINIA : producing only teleutospores which germinate soon after maturity.

It is customary for brevity to designate the aecidia as I, the uredo as II, and the teleutospore as III.

The rusts are grouped in four families of which only two * are known from the United States. These may be separated as follows :

Teleutospores in flattened or cushion-like masses, or loose in the tissues of their host; stemless.   Fam. 1. **Melampsoraceae**.

Teleutospores stalked (rarely the stalk is very short), separate or united in sori of a definite form.   Fam. 2. **Pucciniaceae**.

Family 1. Melampsoraceae.

This family has the following genera in the United States :—

1. Teleutospores formed in rows by successive division of the sporophore, bursting through the tissues of their host.   2.
   Teleutospores not formed in rows; sori remaining covered by the epidermis or cuticle of their host.   3.
2. Teleutospores formed in cushions.   CHRYSOMYXA.
   Teleutospores formed in columnar or filiform masses.   CRONARTIUM.
3. Teleutospores formed in waxy masses, dividing into four cells of which the uppermost bear sterigmata.   COLEOSPORIUM.
   Teleutospores germinating with normal promycelium.   4.
4. Teleutospores one-celled (rarely 2-celled with one cell above the other), uniting laterally in an irregular crust.   MELAMPSORA.
   Teleutospores mostly 2-4-celled, the cells side by side.   5.
5. Teleutospores united in a thick crust formed either in the epidermal cells or immediately under them.   6.
   Teleutospores single or in loosely united groups, buried in the parenchyma of their host; (parasitic on ferns.)   UREDINOPSIS.

---

* Species of the genus *Endophyllum* which produce teleutospores within a pseudoperidium will be found, doubtless, on some of our fleshy plants. The three known European species grow on *Sedum, Sempervivum*, and like plants; they form the type of a distinct family.

6. Sori in limited areas occasionally anastomosing on the leaves.
                                                    PUCCINIASTRUM.
   Sori forming expanded areas mostly on stems.   CALYPTOSPORA.

The genera of this family are mostly composed of few species. *Chrysomyxa* is mainly confined to the Ericaceae; *C. pirolae* is common on species of *Pyrola*. Species of *Cronartium* are found on *Comandra*, *Ribes* and *Quercus*. Species of *Coleosporium* are abundant on various Compositae and are occasional on the common bellwort (*Campanula*). The aecidial forms of the last three genera are members of the form-genus *Peridermium*, which are formed on various conifers. Several of these inhabit the leaves of pines* and one or more species form larger or smaller swellings on the branches or even the trunks of pine trees; we have seen these on pines in Alabama fifteen inches in diameter. Other species of *Peridermium* are found on species of *Abies*.

The species of *Melampsora* are parasitic on willow, poplar and birch, as well as on various herbaceous dicotyledonous genera as *Linum*, *Euphorbia* and *Croton*. The uredo forms are best known but are insufficient for distinguishing the species. The willow-inhabiting species of this country especially need careful study.

Pucciniastrum is best known in this country from the common species parasitic on *Agrimonia* whose uredo form is everywhere common; other species are found on *Prunus* and certain Ericaceae. *Calyptospora* has a single species parasitic on species of *Vaccinium* causing enlargment of the stems; *Uredinopsis* has a few species parasitic on ferns.

Family 2. Pucciniaceae.

This family contains by far the largest number of the rusts, including those that represent parasitic diseases of cultivated plants. Besides the grain-rust belonging to the genus *Puccinia* with several species, we have other species parasitic on *Asparagus*, hollyhocks, corn, sunflower, plum and peach, besides numerous species on weeds and various wild plants of nearly every family. Species of *Uromyces* are parasitic on clover, on beans, on carnations, and on beets, besides numerous species on weeds and vari-

---

* *Cf.* Underwood & Earle. Notes on the pine-inhabiting Species of Peridermium. Bull. Torrey Bot. Club, 23: 400-405. 1896.

ous wild plants. Of the species of *Gymnosporangium*, which produce cedar-apples or other deformities on *Juniperus* we have already spoken ; six species are found on *Juniperus Virginiana* * alone ; other species are found on *J. communis*, *J. occidentalis*, and *Chamaecyparis thyoides*.

The genera found in the United States may be distinguished as follows :

1. Teleutospores imbedded in masses of jelly, mostly 2-celled; (parasitic on Cupressinae). *GYMNOSPORANGIUM.†
Teleutospores in definite sori, not imbedded in jelly. 2.

2. Teleutospores with a simple stalk which is occasionally obsolete. 3.
Teleutospores united in cushion-like bodies formed of several or many cells ; stems formed of several united or separate stalks. RAVENELIA.

3. Teleutospores 1-celled. UROMYCES.†
Teleutospores normally 2-celled. 4.
Teleutospores 3–several-celled. 6.

4. Endospore of teleutospores becoming mucilaginous and swollen.
UROPYXIS.
Endospore of teleutospores without a mucilaginous layer. 5.

5. Aecidia without a pseudoperidium ; spermogones spherical ; (a single species parasitic on *Rubus*). GYMNOCONIA.
Aecidia, when present, with a pseudoperidium, often wanting.
PUCCINIA.†

6. Septa at right angles to the axis of the spore. 7.
Septa 3, uniting in triangles. TRIPHRAGMIUM.

---

* *Cf.* Underwood & Earle. The Distribution of the Species of Gymnosporangium. Bot. Gaz. **22** : 255–258. 1896.

† Among the unfortunate features connected with systematic study are the troublesome questions of synonymy and priority. As in many other things, we of the present generation suffer from the lack of system of the generations behind us. Asa Gray stated succinctly the basis for the difficulty : "For each plant or group there can be only one valid name and that always the most ancient if it is tenable." Now, it is claimed, Micheli in 1729 established the genus *Puccinia* for a gelatinous parasite of the Juniper (= *Gymnosporangium* as used above), and Adanson in 1763 adopted this use of the name long before it was diverted to its present usage. If these are the real facts, *Puccinia* may have to replace *Gymnosporangium* as an older name. In a similar way *Dicaeoma* may have to stand for *Puccinia*, *Caeomurus* for *Uromyces* and *Aregma* for *Phragmidium*. This is

7. Endospore of teleutospore becoming mucilaginous and swollen in water; (parasitic on *Leguminosae*). PHRAGMOPYXIS.
Endospore of teleutospore without a mucilaginous layer; (parasitic on *Rosaceae*). PHRAGMIDIUM.*

Of the above genera *Puccinia* contains about three hundred species within our limits and *Uromyces* about one hundred. *Cropyxis** and *Phragmopyxis* each contain a single species, the former on *Amorpha* and the latter on a Texan leguminous plant of unknown genus. *Gymnoconia* is the cause of the common blackberry or raspberry rust which is everywhere abundant causing whole plants to be covered in spring with the bright orange sori of the aecidial stage of the fungus † *Triphragmium* is a small genus (*Pl. 6. f. 13*), one of the best known species being found on *Aralia nudicaulis* but this does not seem to appear below a certain altitude.

Phragmidium has several species on *Rosa*, *Potentilla* and *Rubus*, the species on cultivated roses being very abundant in some localities and often causing considerable damage (*Pl. 6. f. 14*).

Ravenelia has several American species mostly on members of the Leguminosae (*Pl. 6. f. 11, 12*).

Besides the genera above named there is a large residue of imperfect forms that must remain in the various form-genera until their true relations are known. The greater part of these are cluster-cups belonging to the genus *Aecidium* whose characters have been already set forth (*Pl. 6. f. 18*). Others for the same reason must stand in *Roestelia*. Some aecidial forms without a pseudoperidium belong to the genus *Caeoma*. A few isolated forms represent the uredo stage and these must remain in like manner in the genus *Uredo*. Among these is *Uredo ficus* which causes the rust of the fig.

not the place to introduce novelties in generic nomenclature nor to adopt names that have not been somewhat generally adopted, but it is desirable to call attention to these possible changes which are only part of those that must come when fungus nomenclature is reduced to a rational system and is harmonized with that of the higher plants.

* Dietel in *Die natürlichen Pflanzenfamilien* does not accept this genus while he adopts *Phragmopyxis* which is distinguished by precisely the same characters.

† This fungus has been variously known as *Caeoma nitens* and *Puccinia interstitialis*.

LITERATURE.

Saccardo. Sylloge Fungorum, 7 : 528-882 ; 9 : 291-334 ; 11 : 174-230.

Dietel. Die natürlichen Pflanzenfamilien, 1**: 24-81. 1897.

Winter. Rabenhorst's Kryptogamen Flora Deutschlands u. s. w. 1¹ :

Plowright. A Monograph of the British Uredineae and Ustilagneae, 1-57, 105-271. 1889.

Farlow. The Gymnosporangia or Cedar-apples of the United States. Anniv. Mem. Boston Soc. Nat. Hist. 1-38. *Pl. 1, 2.* 1880

Thaxter. On certain Cultures of Gymnosporangium with Notes on their Roesteliae. Proc. Am. Acad. 22 : 259-269 1887.

Burrill [& Seymour]. Parasitic Fungi of Illinois. Part I. Uredineae. Bull. Ill. State Lab. Nat. Hist. 2 : 141-255. 1885.

Dietel. Die Gattung Ravenelia. Hedwigia, 33 : 22-69. *Pl. 1-5.* 1894.

Arthur & Holway. Uredineae Exsiccatae et Icones. (Current.) Two fascicles of this series of specimens and illustrations have appeared. They are of the greatest value to any one investigating the order.

CHAPTER VIII

THE HIGHER BASIDIOMYCETES *

(Bracket-fungi, Mushrooms and Puffballs.)

Although some of the first orders to be treated here possess very close morphological relationships with those treated in the last chapter, their saprophytic habit links them with the orders to follow. This is especially true of the next order treated.

Order 3. AURICULARIALES.

Two families compose this order distinguished as follows:
Hymenium gymnocarpous (open), ear-shaped or tubercular.
Auriculariaceae.
Hymenium angiocarpous (closed before maturity), more or less globular, stalked. **Pilacraceae.**

The family AURICULARIACEAE with us contains *Auricularia*, the Jew's ear, a gelatinous ear-like fungus growing singly or in clustered masses on *Sambucus*, *Fraxinus* or *Hicoria*, representing one or more species, and *Mylittopsis* an unusual tubercular form that is little known and of uncertain relations. Species of *Auricularia* allied to our own but larger, are used for food by the Chinese.

The family PILACRACEAE contains two genera of which *Pilacre* is represented with us by an inconspicuous capitate fungus growing on beech (*Fagus*) or *Carpinus*.

LITERATURE.

Saccardo. Sylloge Fungorum, 4: 579-581 ; 6: 760-771 ; 11: 142-146.
Lindau. Die natürlichen Pflanzenfamilien, 1**: 82-88. 1897.
Tulasne. Nouvelles notes sur les Fungi Tremellini et leurs alliés. Ann. Sc. Nat. V. 15: 215-235. 1872.
Brefeld. Untersuchungen aus dem Gesammtgebiete der Mykologie, 7: 27-80. *Pl. 1-4*. 1888.

---
* For synopsis of orders treated in this chapter, see pp. 80, 81.

Order 4. TREMELLALES.

This order, which contains the greater portion of the gelatinous fungi, is composed of two small tropical families each containing a single genus, in addition to the widely distributed Tremellaceae. The members of this family are made up of soft watery gelatinous masses ranging in color from white and pink to orange-yellow or black. A common white species often forms extensive masses on old stumps; a common yellow one with brain-like folds is found in smaller masses on hemlock in the Northern States and on dead branches of various sorts in the South. A common black species, *Exidia glandulosa*, forms extensive flattish patches on dead branches. All of these forms dry down to a thin film, but most will revive again when moistened.

The American genera may be distinguished as follows:

1. Forming a thin smoothish crust, with mould-like conidiophores.
       SEBACINA.
   Thick, gelatinous, smooth, plicate or with brain-like convolutions. 2.
   Plicate, with teeth underneath.   TREMELLODON.
   Funnel-shaped with the hymenium inferior.   GYROCEPHALUS.

2. Blackish, smooth or slightly plicate, papillate; conidia* hook-shaped.
       EXIDIA.
   Yellowish-brown, foliaceous; conidia straight, capitate. ULOCOLLA.
   Yellowish or whitish with brain-like convolutions or folds; conidia yeast-like.   TREMELLA.

Ulocolla foliacea forms large pale brownish foliaceous masses growing in the vicinity of stumps. One of the species of *Tremella* occasionally occurs as a parasite on the stalks or pileus of agarics. *Tremellodon* is a clear crystalline structure resembling a *Hydnum;* in fact several of the tremelline forms, particularly those of tropical regions, seem to simulate forms of the higher coriaceous or fleshy species of the Agaricales. Very few of our species are either well known or well characterized, though many of them are quite common. They offer an interesting field for some enthusiastic and careful investigator.

LITERATURE.

Saccardo. Sylloge Fungorum, 6: 772-796; 9: 257-259; 11: 146-149.

---

*These are formed at the germination of the ordinary spores.

Lindau. Die natürlichen Pflanzenfamilien, 1¹**: 88–96. 1897.
Tulasne. Observations sur l'organisation des Trémellinées. Ann. Sc. Nat. III. **19**: 193–231. *Pl. 10–13.* 1853.
Brefeld. Untersuchungen aus dem Gesammtgebiete der Mykologie, **7**: 80–138. *Pl. 5–8.* 1888.

Order 5. DACRYOMYCETALES.

This order is quite closely allied to the last in habit, but differs in the undivided basidia which ally it more closely with the higher forms of Agaricales. The species are mostly small and comparatively inconspicuous, some are flat and effused, others are elongate and resemble flattened clubs.

The American genera may be distinguished as follows:

1. Sessile, depressed or resupinate.	2.
Elongate or with a distinct stipe.	4.
2. Mould-like with a waxy smoothish hymenium.	ARRHYTIDIA.
Thin, forming a waxy effused crust.	CERACEA.
Cushion-shaped or globular, gelatinous.	3.
3. Conidia globular.	HORMOMYCES.
Conidia elliptic.	DACRYOMYCES.
4. Hymenium covering only a portion of the fungus.	5.
Hymenium extending to all parts of the fungus.	6.
5. Cartilaginous or fleshy, capitate.	DITIOLA.
Gelatinous or fleshy when dry, spatulate or goblet-shaped.	GUEPINIA.
6. Capitate; spores ultimately septate.	DACRYOPSIS.
Capitate; spores simple, one on each basidium.	COLLYRIA.
Awl-shaped, simple or branched.	CALOCERA.

The species of *Calocera* resemble small species of the club fungi (*Clavaria*). *Guepinia* has a common yellow species that grows in the cracks of rails and logs, forming slender spatulate forms a half an inch or more high. Like the members of the preceding order, the species are poorly known.

LITERATURE.

Saccardo. Sylloge Fungorum, **6**: 796–815; **9**: 259–261; **11**: 149–151.
Hennings. Die natürlichen Pflanzenfamilien, 1¹**: 96–102. 1898.

Tulasne. Observations sur l'organization des Trémellinées. Ann. Sc. Nat. III. **19**: 193–231. *Pl. 10–13.* 1853.

Tulasne. Nouvelles notes sur les Fungi Tremellini et leurs alliés, Ann. Sc. Nat. V. **15**: 215–235. 1872.

Brefeld. Untersuchungen aus dem Gesammtgebiete der Mykologie, **7**: 138–167. *Pl. 10, 11.* 1888.

Order 6. EXOBASIDIALES.

This order is composed of a single family, the Exobasidiaceae, and contains parasitic species which produce deformities and galls chiefly on members of the *Ericaceae*, that are superficially comparable to the deformities produced by species of *Exoascus* on the plums. One of the common species produces large swollen whitish or pinkish galls on *Azalea* which have a slightly acid taste and are sometimes eaten under the name of May apples; others produce disk-like deformities on leaves of the same plant. Others still are found on *Cassandra*, cranberrries and various species of *Vaccinium* and *Gaylussacia*, either deforming the leaves, in which pockets are formed, or the flowers and young fruit, which become enormously swollen and assume a pinkish white tint. Certain cultures have been made that seem to indicate that several of these various forms may be members of a single polymorphic species.

Besides the genus *Exobasidium* with four-spored basidia, a second genus, *Microstroma*, has recently been placed in this family, distinguished by its mostly six-spored basidia. Two species are known, one growing on oak leaves and the other common on *Juglans* and *Hicoria* forming whitish patches on the under surface of the leaves.

LITERATURE.

Hennings. Die natürlichen Pflanzenfamilien, **1****: 103–105.

Saccardo. Sylloge Fungorum, **6**: 664–666; **9**: 244, 245; **11**: 130, 131.

Richards. Notes on Cultures of Exobasidium andromedae and of Exobasidium vacinii. Bot. Gaz. **21**: 101–108. *Pl. 6.* 1896.

Order 7. AGARICALES.

This extensive order has been commonly known under the name HYMENOMYCETES, the name arising from the fact that the basidia which bear the spores are arranged in a sort of membrane (*hymenium*) which covers various exposed portions of the fungus either the ends of clubs or coral-like branches as in *Clavaria*, or spread out over a flattish surface as in *Stereum*, or covering the surface of spines, teeth or thin plates or lamellae as in *Hydnum* and *Agaricus*, or lining the interior surface of pores or tubes as in *Boletus* and *Polyporus*. In this order generic and family distinctions are often confused since exceptional forms stand on the boundary lines of two genera or even of two families. The following families, however, in most cases, may be somewhat easily distinguished :

1. Club-shaped or forming masses of erect branches rising from a common base ; spores borne on the upper portions. Family 3. **Clavariaceae**.
Provided with a cap (pileus) and central stem, or bracket-like, or entirely resupinate ; spore-bearing surface normally underneath.   2.

2. Spores borne on radiating lamellae.   Family 7. **Agaricaceae**.
Spores borne on teeth, tubercles, or tooth-like plates,* projecting downward in growth.   Family 4. **Hydnaceae**.
Spores borne on the interior of pores or tubes or labyrinthine passages,† rarely with merely shallow anastomosing folds.   3.
Spores covering smooth or mould-like surfaces, only slightly roughened or wrinkled.   4.

3. Pores more or less readily separating from the pileus and from each other ; substance fleshy ; stem central or lateral.
Family 6. **Boletaceae**.
Pores not easily separating from the pileus ; substance rarely fleshy, more commonly leathery, corky or woody. Family 5. **Polyporaceae**.

4. Basidia loosely aggregated on a mould-like or arachnoid base, formed from loose floccose hyphae.   Family 1. **Hypochnaceae**.
Basidia closely aggregated, forming a compact crust-like layer ; resupinate or sometimes pileate.   Family 2. **Thelephoraceae**.

---

* Some forms of Polyporaceae when old have lacerated pores which are often indistinguishable from flattened teeth. This is especially noticeable in *P. pergamenus*, one of our commonest and most widely distributed species.

† In a few cases merely anastomosing lamellae, thus forming an easy transition to certain Agaricaceae.

Of the above families, the Agaricaceae contain the plants ordinarily known as toadstools or mushrooms which are usually provided with a central stem; a few growing from the sides of stumps or logs are stemless, the larger of which are often known as oyster mushrooms.

The Boletaceae form cushion-like fleshy toadstools provided with pores; the beefsteak fungus with a lateral stem also belongs with this family.

The Polyporaceae include the woody, corky and leathery fungi with pores which shelve out from standing tree trunks or fallen logs and are commonly known as bracket fungi. A few have a central stem, more are dimidiate or semicircular in outline attached by the side, while others are resupinate, *i. e.*, have no pileus, but the pores are attached directly to a thin crust like expansion of the ground work or context of the fungus.

The Hydnaceae include forms of fungi with spines, flattened teeth, or irregular tubercles. Some form coral-like masses hanging from the sides of stumps or tree trunks, some have the forms of fleshy agarics with central stems, a few are dimidiate, but the greater part are resupinate, and many are inconspicuous.

The Thelephoraceae include a few pileate forms, and a few, particularly species of *Thelephora*, are terrestrial; the greater part, however, form resupinate crusts of various colors from brown to blue, yellow and white, growing on the sides of standing trunks, or on small branches or under fallen logs. This brings us to the simplest group of all.

Family 1. Hypochnaceae.

This family is not largely represented in either genera or species and seems to form the nearest approach of this order to the Dematiaceae (p. 76) since the foundation for the basidial membrane consists of loose floccose or arachnoid hyphae and the basidia themselves are loosely aggregated. Two of the six genera are found with us and may be distinguished as follows:

Spores uncolored, smooth; basidia with 2–4 (rarely 6) sterigmata.
                      HYPOCHNUS.
Spores colored, mostly spiny.
                        TOMENTELLA.

Two species of *Hypochnus* are common in the Southern States growing on trees and logs, *H. rubro-cinctus* with brilliant scarlet borders, and *H. albo-cinctus* glistening white; other species are less common farther north. The species of *Tomentella* form yellowish-brown crusts on old stumps and logs.

Family 2. **Thelephoraceae.**

This family contains an enormous number of species whose relations are not clearly known. In certain genera enlarged cells known as *cystidia* are found among the basidia often difficult to make out in weathered specimens; in others these take the form of bristles which are persistent even when the hymenial surface is old, and can be seen with a lens or more readily under a low power of the microscope. In some genera the basidia are attached directly to the mycelial layer; in those more differentiated an intermediate fibrous layer is developed between. We have representatives of the following genera:

1. Hymenium without cystidia. 2.
   Hymenium roughened with bristle-like cystidia. 8.
2. Resupinate, with no intermediate fibrous layer. 3.
   Ascending, pileate or stalked (rarely wholly resupinate). 5.
3. Spores colorless throughout. 4.
   Spore contents colored, membrane uncolored. ALEURODISCUS.
   Spore membrane yellowish-brown. CONIOPHORA.
4. Spores sessile. CORTICIUM.
   Spores stalked. MICHENERA.
5. Context formed of different layers, leathery or woody; hymenium mostly smooth; spores colorless. STEREUM.
   Context of a single mostly uniform layer. 6.
6. Coriaceous, pileate or branched; hymenium smooth or slightly warty. THELEPHORA.
   Fleshy (smoothish); terrestrial or rarely epixylous, cup-shaped or umbrella-shaped, stalked. CRATERELLUS.*
   Membranous, cylindric, tubular or cup-shaped; mostly small, epixylous. 7.
7. Growing singly, mostly cup-shaped with flaring sides. CYPHELLA.
   Growing in dense clusters, mostly short tubular. SOLENIA. †

---

* Peck (Bull. N. Y. Museum 1: 44–48) gives a synopsis and descriptions of the five species growing in New York.

† Saccardo and others place this genus among the Polyporaceae. The species externally resemble small species of *Peziza*.

8. Cystidia simple, unbranched. 9.
   Cystidia branched, star-shaped. ASTEROSTROMA.
9. Context uniform, always resupinate. PENIOPHORA.
   Context formed of different layers, often pileate. HYMENOCHAETE.

Of the above genera *Aleurodiscus* and *Michenera* are small genera, the former with two closely allied species often seen on the bark of *Ostrya* where they form discrete patches; the latter genus has only a single species with us. *Coniophora* and especially *Corticium* are larger genera, and our species are poorly known and many of them difficult to identify; probably over one hundred nominal species are described from the United States.

Stereum and *Hymenochaete* represent the common leathery or woody genera growing on logs. The species of the latter genus are usually brownish in color, in addition to the possession of cystidia which are usually easy to discover. *Stereum versicolor* is a common pileate species and *S. frustulosum* is a perennial resupinate species everywhere common on oak whose wood it renders very hard and brittle. Several species of each genus are common and widely distributed. One anomalous species of *Hymenochaete* (*H. agglutinans*) often grows on living twigs and strangles them to death.

Thelephora is more commonly terrestrial and pileate, including leathery species. While most of the species resemble *Stereum* in shape, *T. Schweinitzii* is much branched and is quite frequently mistaken for a *Clavaria;* some other species are even more finely laciniate; *T. Willeyi* is goblet-shaped. One anomalous species (*T. pedicellata*) grows around the living stems of young trees.

Craterellus contains chiefly fleshy species and forms a striking connecting link to the Agaricaceae through *Cantharellus;* so close is the connection that it is difficult in the case of certain specimens to determine whether the plant belongs to one genus or the other. This is only another indication that the limitation groups we are here discussing are artificial rather than natural.

Family 3. Clavariaceae.

The more conspicuous members of this family are club-like and simple or form fleshy coral-like masses of various shades of white, yellow or even brighter colors like pink and violet. A number are less conspicuous and consequently less known, forming waxy or horny, simple or branched bodies. In all cases the members of the family may be distinguished by having the hymenial layer normally apical and exposed, rather than underneath and protected as in the allied families. The genera may be distinguished as follows :

1. Plant small, simple.   2.
   Plant mostly larger, conspicuous, usually branched, but occasionally simple and club-like.   4.
2. Plant capitate, hollow.   PHYSALACRIA.
   Plant clavate or filiform.   3.
3. Basidia with two sterigmata.   PISTILLARIA.
   Basidia with four sterigmata.   TYPHULA.
4. Branches strongly flattened, leaf-like.   SPARASSIS.
   Branches or clubs terete or only slightly compressed.   5.
5. Context fleshy; simple or commonly much branched.   CLAVARIA.*
   Context cartilaginous, horny when dry ; mostly slender filiform.
       PTERULA.
   Context coriaceous ; surface tomentose.   LACHNOCLADIUM.

Sparassis crispa sometimes forms masses as large as one's head ; *S. Herbstii* is a handsome species recently described.

In collecting species of *Clavaria*, the largest genus, ample field notes should be taken, including color, color of spores,† taste, habitat and the character of the apices of the branches. Many of the largest species of *Clavaria* and *Sparassis* are edible and none are known to be deleterious.

Family 4. Hydnaceae.

The members of this family are known as the prickly fungi since the more typical forms are provided with teeth. In the typical genus, *Hydnum*, these teeth are normally terete, but in *Irpex* they become flattened, and in *Radulum* and *Phlebia* they

---

* Morgan (Jour. Cincinnati Soc. Nat. Hist. 11 : 86–90) gives descriptions of twenty species growing in the Ohio Valley, and Peck (Reg. Rep. 24 : 104, 105) gives synopses of twenty New York species without descriptions.

† Easily obtained by laying the plant under a tumbler or bell-jar on paper, preferably colored.

become reduced to mere tubercles or irregular ridges. The genera may be distinguished by the following synopsis :

1. Plant consisting of teeth only with no basal membrane or context.
   MUCRONELLA.
   Plant consisting of teeth attached to a basal membrane or to a pileus 2.
2. Hymenium covering needle-like spines. 3.
   Hymenium covering lamella-like teeth. 4.
   Resupinate, the hymenium with warts, wrinkles or simple bristles. 5.
3. Spines irregular, thick and blunt. RADULUM.
   Spines awl-shaped, usually regular. HYDNUM.
4. Context coriaceous, pileate or resupinate ; epixylous. IRPEX.
   Context fleshy or membranous, pileate ; terrestrial. SISTOTREMA.
5. Hymenial surface with simple bristles. PYCNODON.*
   Hymenial surface with low crested wrinkles. PHLEBIA.
   Hymenial surface warty. 6.
6. Warts hemispheric, smooth. GRANDINIA.
   Warts crested, papillose. ODONTIUM.

Of the above genera *Hydnum* is the only large genus and is composed of numerous groups of diverse habit. Some of these are so distinct that they have been set apart as genera, some of which will doubtless be regarded valid. Several of the species of *Hydnum* are edible. Among them are the following :

1. *Species with a central stem.*

H. repandum has a yellowish or slightly reddish pileus with an irregular margin, is compact but brittle, with a dry, whitish context; the pileus is one to four inches wide and the stem is one to three inches in length.

H. imbricatum is a larger species, often six inches or more in diameter with a brownish pileus covered more or less closely with irregular scales of a darker color ; the centre is sometimes depressed so that the fungus appears funnel-shaped.

2. *Species growing in branched masses with no distinct pileus.*

H. coralloides is a pure white fungus of coral-like character, often composed of many spreading and often interlacing branches covered with short spines scarcely a third of an inch in length. It is a beautiful species most common on trunks of beech wood.

---

*This is the *Kneiffia* of Fries, a name which is preoccupied. *Cf.* Bull. Torrey Bot. Club, **25**: 630, 631. 1898.

H. echinaceus is white, becoming yellowish in the form of a tubercular mass often six to eight inches broad with straight equal spines which are sometimes nearly two inches long.

H. caput-medusae is also tubercular, but changes from white to a smoky or ashy tint and is usually contracted into a stalk behind; it has short distorted teeth above and long uniform ones below.

A large number of the species of *Hydnum* are resupinate, some of them growing in the form of thin layers of mycelium expanded into a membrane on which a few spines are borne; others form larger areas several feet in extent, growing underneath fallen logs.*

Family 5. **Polyporaceae.**

This family contains the pore-bearing forms whose pores are usually permanently united to the context† and to each other. A few of the members of the family are fleshy, some, indeed, being edible, but by far the greater part are leathery, corky, membranous or woody. They form the more conspicuous bracket-fungi shelving from dead or dying trunks and logs, some of them attaining a very large size; others growing in similar situations are very small or even minute. Nearly six hundred species have been reported from America.

The genera can be distinguished as follows:

1. Pores minute and round or larger and angular.   2.
   Pores forming labrinthine passages or becoming lamella-like plates. 8.
   Pores reduced to shallow pits separated by narrow ridges, folds or
      reticulations; more commonly resupinate.   MERULIUS.

---

* Fries used the following subdivisions for many of his genera of which *Hydnum* possesses common examples of all. A statement of their characters will show the diversity of the genus as here limited:

   I. MESOPUS: with a central stem supporting a pileus.

   II. PLEUROPUS: with a lateral or eccentric stem supporting a pileus.

   III. MERISMA: compound or multiple forms united to one base.

   IV. APUS: pileate, semicircular, sessile.

   V. RESUPINATI: without a pileus, the slender membranous base attached flatly to the hymenium.

   † In *Gloeoporus* the layer of pores sometimes peels off from the context when rather young and moist.

2. Normally pileate, only accidentally resupinate. 3.
   Normally resupinate. 6.
3. Pores usually small or medium size, round or angular. 4.
   Pores large, hexagonal or lamelliform. 7.
4. Hymenium waxy, separating like a membrane from the context when wet. GLOEOPORUS.
   Hymenium not waxy; not separable from the context. 5.
5. Trama (*i. e.*, the substance of the pileus) descending between the pores.* TRAMETES.
   Trama not descending; substance between the pores different from that of the pileus. POLYPORUS. †
6. Pores parallel, seated on a membranous or more or less fleshy base. PORIA. ‡
   Pores in the summits of small papillae, fixed to a membranous base. POROTHELIUM.
7. Stems lateral; pores long hexagonal. FAVOLUS.
   Sessile or resupinate; pores regular hexagons. HEXAGONIA.
8. Lamellae concentric: stem central or nearly so. CYCLOMYCES.
   Lamellae radial; sessile or accidentally resupinate. LENZITES. §
   Lamellae sinuous and labyrinthine; normally sessile. DAEDALEA.

Several of the above species have few representatives in our flora. *Gloeoporus* has only one, *G. conchoides*, which is comparatively common; *Favolus* has one species everywhere common on beech and hickory and two or three less common; *Cyclomyces* has a single extremely rare species; *Hexagonia* has only one or two species found only southward; *Porothelium* has only one species that is anywhere common. *Lenzites* and *Daedalea* have each a dozen or more species of quite diverse habit, and *Me-*

---

* This distinction is often very poorly defined in many species and shows an artificial separation of genera.

† Saccardo and others separate from this genus 1. *Fomes*, woody species with mostly stratose pores, and 2. *Polystictus* with leathery context, leaving the more or less fleshy species in *Polyporus*. These distinctions at best are artificial; Karsten has distinguished several genera.

‡ Almost any species of *Polyporus* may become resupinate and this genus doubtless contains several described species that are normally pileate.

§ By many this genus has been placed among the Agaricaceae because of its lamellae which only occasionally anastomose. Its habit places it here and it really forms a connecting link between the two families.

rulius and *Trametes* have thirty or forty species each ; *M. tremellosus* is a semi-fleshy species which is common under old logs in the latter part of the season ; *Trametes suaveolens* is a white species common on willows; *T. cinnabarina* is bright red, growing commonly or birch and cherry, while *T. pini* is a brownish species growing on pines and other coniferous trees.

The genus *Polyporus*, in its widest sense including *Poria*, contains nearly five hundred American species many of which are poorly known and defined. Several attempts have been made to separate the enormous genus into natural groups or genera some of which are really well defined, but until a revision of the entire genus can be made it is as well to allow the species to rest under a single name. Many of the genera which have been proposed have resulted in associating together very unlike plants.*

A few of the fleshy species are edible. Notable among them is *P. sulfureus*, which often grows in prodigious masses of overlapping pilei; it is pinkish yellow above and the pores are a bright sulphur yellow beneath. Of course, it is edible only when young, about the time that the pores commence to develop, since the plant toughens with age. Certain species of *Polyporus* appear to be confined to certain trees while others seem to be independent of their substratum and are likely to appear on any old log. Some of the fleshy or semi-fleshy species soon decay while some of the stratose species live many years forming layer after layer of pores.

Family 6. Boletaceae.

This family includes by far the greater number of perishable fleshy fungi with pores, comprising those in which the pores quite easily separate from the pileus and from each other. In one genus, *Strobilomyces*, this is less marked, and this genus forms a rather natural transition to some of the fleshy species of *Polyporus*. On the other hand, *Boletinus*, in which some of the species have the pores arranged in radiating lines, forms a somewhat easy transition to *Paxillus* among the agarics.

---

* *E. g., Mucronoporus*, which has been proposed to include those species in which spines (cystidia) are scattered among the basidia lining the pores. *Myriadoporus* is probably founded on a deformed species of *Polyporus adustus*.

There are four * genera in this country distinguished as follows:

1. Stem strictly lateral; pores in the form of tubes whose mouths are separated from each other. FISTULINA.
   Stem central or rarely somewhat eccentric. 2.
2. Pores readily separating from the pileus; spores brownish or whitish.
   BOLETUS.
   Pores separating with difficulty from the pileus. 3.
3. Fleshy; pores in more or less radiating rows; spores brown or yellowish.
   BOLETINUS.
   Tough; pores uniform; pileus floccose; spores blackish
   STROBILOMYCES.

Strobilomyces has a single species everywhere common which is grayish or blackish and easily recognized by its floccose pileus; it is edible when young, but soon becomes tough; a second species is imperfectly known. *Boletinus* has a few species one of which, *B. porosus*, is readily recognized by its eccentric stem and lamella-like radiating pores; *B. pictus* is a pinkish species with a conspicuous membranous veil. *Fistulina hepatica* is known as vegetable beefsteak from the flesh-like fibre and color of the fresh specimen; it is comparatively common in autumn on chestnut stumps and trunks, sometimes also growing on oak; it is much valued as a delicacy.

The species of *Boletus* are quite diverse and we have over a hundred of them.† An entire group known as the *Luridi* from the red mouths of the pores are suspected of being poisonous and should be avoided. Very recently a case of poisoning has been reported from a species outside of this group.‡ Beyond these

---

* Hennings (Die natürlichen Pflanzenfamilien, 1^{**} : 188–196) separates from *Boletus* two additional genera which are also represented in this country.

† Fortunately descriptions of the greater number of the species of this genus are easily accessible in English. *Cf.* Peck, Boleti of the United States. Bull. N. Y. State Mus., No. 8, 1888. This can be obtained at a small price of the State Librarian at Albany, N. Y. Quite a number of species have been described since this publication was issued, particularly from the South.

‡ *Boletus miniato-olivaceus sensibilis*, *Cf.* F. S. Collins in Rhodora, 1 : 21–23. 1899. This is the species erroneously figured in Palmer's Mushrooms of America as *B. subtomentosus* and edible.

none of the species are known to be injurious except *B. felleus*, a common species with brown pileus, reticulated stem, and flesh-colored pores, whose bitter taste would prevent it from being used as food. In this genus as everywhere among the fungi, unknown species should be tested with caution and special care should be taken to avoid specimens that are not strictly fresh and young. Among the edible species we may mention the following as well recommended:

1. *Boleti with a viscid pileus.*

B. granulatus is usually grayish yellow with pale yellow pores and white flesh; the mouths of the tubes are dotted with minute brownish granules from which the specific name is derived.

B. luteus has a yellowish-brown pileus, with minute yellow pores and whitish or slightly yellowish flesh; the stem has an annulus above which it is yellowish with small brown dots. It is not a common species. The pileus is from two to five inches across and the stem is scarcely two inches long.

B. subluteus has a dingy yellowish-brown pileus and with a whitish or dingy yellowish stem which is marked by brown dots both above and below the annulus. It is slightly smaller than the last-named species and has a more slender stem.

2. *Boleti with a dry pileus.*

B. scaber is easily recognized by its whitish pores and its long stem which is roughened by small brownish or blackish scales. The pileus is variable in color, ranging from nearly white to almost black, and is from two to five inches across with a stem from three to six inches long. It is a very common species.

B. edulis has a grayish red or brownish pileus, and its whitish flesh is tinged with reddish just beneath the cuticle; the pores are whitish at first and become greenish yellow with age. The pileus is four to six inches across and the stem is from two to six inches long, and is marked by a network of raised lines just beneath the layer of pores.

B. castaneus has a dull reddish or cinnamon-colored pileus, shallow pores which are whitish at first and then yellowish, and a hollow stem in color not unlike the pileus.

B. versipellis has a yellowish-red pileus two to six inches across, long small pores which are grayish-white becoming dingy with age; its veil frequently clings in torn fragments to the margins of the pileus; the stem is solid and of a whitish color and is marked somewhat like that of *B. scaber*. Several species of *Boletus* are canned in Germany and sold under the name of *Steinpilze*.

Some of the species of *Boletus* are very beautiful. A very common yellow species (*B. ornatipes*) has the stem elegantly reticulated. *B. Ravenelii*, more common in the Southern States, is a brilliant yellow and is dusted over with yellow powder. *B. Frostii*, of the section *Luridi* has an elegant red and yellow mottled pileus and the tubes have the characteristic bright red mouths of the section. It is a common species of eastern New York and New England, and probably has a wide distribution. *B. auriporus* has a pileus and brilliant golden yellow pores. Many of the species change the color of the flesh to a brilliant blue on exposure to the air; while this is not a sure sign of an injurious species, it is to be looked upon as a suspicious character.

The species of *Boletus* are most abundant during the moist rains of late summer and early autumn, ranging in the Northern States from July to October and in the Southern States appearing somewhat earlier if there is sufficient soil moisture, as well as somewhat later than farther northward.

Family 7. Agaricaceae.

This family is the largest and most widely distributed of all the families, containing some five thousand described species. Over twelve hundred are known from Great Britain alone, and about as many have been reported from the United States, where careful, exhaustive research has been made only in a few limited areas and by a very few individuals.

The common edible mushroom of the fields and markets, *Agaricus campestris*, is a type of this family and a somewhat detailed account of its character may be given with side references to structures illustrated by species of allied genera. The growing or vegetative portion of the fungus consists of wide spreading hyphae forming a tangled mycelium which permeates the soil in search of decomposing organic matter which serves as food to the plant. This mycelium, grown in compost and dried, forms the so-called spawn of the seed dealers from which the ordinary cultivated mushrooms are propagated.* From this mycelium the young mushrooms form as mere rounded masses, later taking on the form of the buttons that are familiar to us in the ordinary French canned mushrooms ; as they continue to emerge above the surface of the soil they take on the umbrella-like form so familiar to us, the spores mature on the hymenial surface which covers the lamellae or so-called gills, soon after which the mushroom decays, the latter process too frequently hastened through the agency of larvae, which often attack certain species of mushrooms before they are mature. An ordinary mushroom consists of two parts, the pileus or cap and the stalk or stipe (*Pl. S. f. 1*). In some genera a veil is present which in the young stage extends across from the stipe to the margin of the pileus ; as the pileus expands, the veil ruptures and either remains as a collar or ring (*annulus*) about the stipe (*Pl. S. f. 1, a*) or hangs in tatters as a fringe at the margin of the pileus. In the genus *Cortinarius*, the veil is arachnoid like a delicate cobweb and, as the expansion of the pileus takes place, collapses and leaves little trace of its existence. In a number of genera there is also a volva or universal veil which is attached at the base of the stipe and envelopes the entire mushroom when young (*Pl. S. f. 2.*) As the mushroom expands, this volva ruptures and is either carried up in a series of flocculent scales on the pileus as in the fly agaric (*Amanita muscaria*) or a part or all of it remains as a permanent cup at the base of the stipe as in the deadly *Amanita phalloides* (*Pl. S. f. 1, b*). The characters on which genera and species are based in addition to color and habit are the color of the spores, taste and odor, the position of the lamellae, their shape and their character, the nature of the stem and its relation to the context or flesh of the pileus and various other characters that will be noted under the synopsis of genera.

The color of the spores is easily determined by cutting a pileus

---

* The literature devoted to the cultivation of mushrooms is quite extensive. Among the best works treating of the subject are :

Falconer. Mushrooms ; how to grow them. 1896.

Robinson. Mushroom Culture; its Extension and Improvement. 1870.

One of the Farmer's Bulletins (No. 53), published for gratuitous distribution by the Department of Agriculture at Washington, gives ample directions for ordinary cultivation.

from the stem and placing it with the lamellae downward on a sheet of paper under a bell jar or tumbler. In from two to six hours, according to the maturity of the specimen, the spores will fall in radiating lines from the lamellae and produce a spore print.* If a microscope is at hand the form and sizes of the spores should also be noted.

The taste and odor of many species are peculiar. Certain species have a nutty or mealy flavor, some are bitter and disagreeable, some have a biting peppery taste. It is perfectly safe to sample any species by directly tasting a small fragment, the only discomfort that can arise being a lingering taste similar to that coming from red pepper which is found in a few species.

The lamellae may be free from the stem (*Pl. 7. f. 3*) or attached, in which case they may be merely adnate (*Pl. 7. f. 1*), or sinuate (*Pl. 7. f. 2*), or decurrent (*Pl. 7. f. 4*). It should also be noted whether the lamellae are uniform or alternate with a series of shorter ones (heterophyllous) ; in a few cases the lamellae are connected by cross veins or septa.

The pileus itself varies in shape from oval and narrowly conic to widely conic and umbrella-like, or in some cases the edges turn upwards at maturity like an umbrella turned inside out. The surface may be dry, watery (hygrophanous), or viscid and either smooth, scaly, mealy or otherwise ornamented. The disc or central part of the pileus often presents special characters ; besides being sometimes of a different color than the rest of the pileus, it is sometimes depressed (umbilicate) (*Pl. 7. f. 5*), or even funnel-like (infundibuliform) (*Pl. 7. f. 6*), or raised into a rounded prominence (umbonate) (*Pl. 7. f. 4*).

The stipe may be fleshy and continuous with the context of the pileus or may be cartilaginous. The interior may be hollow (*Pl. 7. f. 2*), solid, or filled with a loose mass of hyphae (stuffed).

The genera of the family are quite numerous and except in a few abnormal cases of variation may be distinguished as follows :

1. Plant fleshy, soon putrescent. 2.
   Plant tough, leathery or woody, reviving or persistent. 13.

---

* These spore prints may be made permanent by spraying them from an atomizer with a solution of white shellac in alcohol. A saturated solution should be made and then diluted fifty per cent. with alcohol. If the spores are white the print should be taken on colored paper.

2. Juice milky, white or colored.   LACTARIUS.
    Juice watery.   3.
3. Stem lateral, eccentric or wanting.   4.
    Stem central, or nearly so.   5.
4. Spores white (violet tinted in one species).   PLEUROTUS.
    Spores rosy or salmon-colored.   CLAUDOPUS.
    Spores yellowish brown.   CREPIDOTUS.
5. Spores white (green in *Lepiota Morgani*).   6.
    Spores rosy or salmon colored.   16.
    Spores yellowish brown or rusty brown.   19.
    Spores dark brown or purplish brown.   24.
    Spores black.   28.

(White-spored Series.)

6. With a volva* and annulus.   AMANITA.
    With a volva but no annulus.   AMANITOPSIS.
    Volva wanting ; annulus present.   7.
    Both volva and annulus wanting.   8.
7. Lamellae free from the stem ; annulus often moveable ; pileus usually scaly, sometimes densely so.   LEPIOTA.
    Lamellae united with the stem ; pileus usually smooth (often somewhat scaly in *A. mellea*, a common species).   ARMILLARIA.
8. Lamellae thin, their edges acute.   9.
    Lamellae in the form of shallow folds, their edges obtuse.   12.
9. Lamellae decurrent on the stem ; stem fleshy.†   CLITOCYBE.
    — stem with cartilaginous rind.   OMPHALIA.
    Lamellae adnate ; stem with a cartilaginous rind.   COLLYBIA.
    — stem fleshy ; pileus often bright colored.   10.
    Lamellae sinuate ; stem fleshy.   TRICHOLOMA.
    — stem with a cartilaginous rind.   11.
10. Plant rigid, the lamellae usually brittle.   RUSSULA.
    Plant with waxy lamellae.   HYGROPHORUS.
11. Pileus membranous, more or less striate.   MYCENA.
    Pileus very thin, without pellicle.   HIATULA.

---

* The volva will appear either as a cup at the base of the stem, or as separable floccose scales on the pileus.

† By cutting the pileus longitudinally through the centre of the stem this feature will be apparent ; in species with a fleshy stem the flesh is continuous with the context of the pileus.

AGARICALES

12. Lamellae decurrent ; plant terrestrial.   CANTHARELLUS.
    Lamellae adnate ; plant parasitic on other Agarics.   NYCTALIS.
13. Lamellae normally toothed on their edges ; stem central, eccentric or lateral.   LENTINUS. *
    Lamellae entire ; stems central.   14.
        —stems lateral or wanting.   15.
14. Lamellae simple ; pileus firm and dry.   MARASMIUS.
        —pileus somewhat gelatinous.   HELIOMYCES.
    Lamellae branched.   XEROTUS.
15. Lamellae simple ; plant leathery.   PANUS. *
    Lamellae deeply splitting, villous.   SCHIZOPHYLLUM.
    Lamellae channeled or crisped, smooth.   TROGIA.

(Pink-spored Series.)

16. Volva present ; annulus wanting.   VOLVARIA.
    Volva wanting ; annulus present.   ANNULARIA.
    With neither volva nor annulus.   17.
17. Lamellae free from the stem.   PLUTEUS.
    Lamellae adnate or sinuate ; stem fleshy.   ENTOLOMA.
       — stem with a cartilaginous rind.   18.
    Lamellae decurrent on the stem ; stem fleshy.   CLITOPILUS.
       — stem with a cartilaginous rind.   ECCILIA.
18. Pileus torn into scales.   LEPTONIA.
    Pileus papillose, subcampanulate.   NOLANEA.

(Rusty-spored Series.)

19. Annulus continuous   PHOLIOTA.
    Annulus arachnoid, filamentous or evanescent, often not apparent in older specimens.   20.
    Annulus wanting.   21.
20. Lamellae adnate ; plants terrestrial.   CORTINARIUS.
    Lamellae decurrent ; plants mostly epipyhtal.   FLAMMULA.
    Lamellae deliquescent, almost separate from the stem.   BOLBITIUS.
21. Lamellae decurrent, easily separating ; stem fleshy.   PAXILLUS.
        —stem with a cartilaginous rind.   TUBARIA.
    Lamellae not decurrent ; stem fleshy.   22.
        —stem with a cartilaginous rind.   23.

---

* Some species of *Lentinus* with entire gills can scarcely be distinguished from *Panus* ; some of the more fleshy forms of the latter are quite close to some forms of *Pleurotus* ; *Panus* and *Lentinus* are sometimes united in one genus.

22. Pileus fibrillose or silky.	INOCYBE.
Pileus smooth and viscid.	HEBELOMA.
23. Margin of pileus incurved when young.	NAUCORIA.
Margin of pileus straight ; pileus viscid ; lamellae free.	PLUTEOLUS.
— pileus not viscid ; lamellae attached.	GALERA.

(Brown-spored Series.)

24. With a volva at the base.	CHITONIA.
Without a volva.	25.
25. Veil remaining on the stem as an annulus.	26.
Veil remaining attached to the margin of the pileus, often not apparent in very old specimens.	HYPHOLOMA.
Veil inconspicuous or wanting ; lamellae free.	PILOSACE.
— lamellae decurrent.	DECONICA.
— lamellae adnate or sinuate.	27.
26. Lamellae free from the stem.	AGARICUS.
Lamellae united with the stem.	STROPHARIA.
27. Margin of pileus incurved when young.	PSILOCYBE.
Margin of pileus always straight.	PSATHYRA.

(Black-spored Series.)

28. Stem dilated above into a disc which bears the radiating lamellae.	MONTAGNITES.
Pileus of the normal form, leathery or horny.	ANTHRACOPHYLLUM.
— fleshy, membranous or deliquescent.	29.
29. Lamellae deliquescent, melting to an inky fluid.	COPRINUS.
Lamellae not deliquescent ; annulus present.	ANELLARIA.
— annulus wanting.	30.
30. Lamellae waxy, decurrent ; spores fusiform.	GOMPHIDIUS.
Lamellae not waxy nor decurrent ; spores globose-ovoid.	31.
31. Pileus striate ; stem with a cartilaginous rind.	PSATHYRELLA.
Pileus not striate ; stem fleshy.	PANAEOLUS.

The above arrangement of the genera, like every other which is based on single characters, is clearly artificial. Hennings[*] has given a recent attempt at a natural arrangement, and while we cannot agree with his combinations of genera nor with his shifting of several generic names, we present an outline of his main groups for the discussion of individual genera.

---

[*] Die natürlichen Pflanzenfamilien, 1**: 198-276. 1898.

1. *Species with obtuse fold-like lamellae* (CANTHARELLEAE).

Trogia has a single small coriaceous species with brownish pileus and white hymenium growing commonly on fallen branches of alder.

Cantharellus contains the fleshy speces of this group, of which the egg-yellow *C. cantharellus* is the best known. It is very common in the forests of Germany and other parts of Europe whence it is commonly carried to the markets, as well as in America. *C. cinnabarinus* is a smaller bright orange reddish species.*

To this section the species of *Craterellus* (p. 100) will be added when we attain to a natural classification of these plants.

2. *Fleshy species with anastomosing lamellae* (PAXILLEAE).

Paxillus is the only genus of this section. Of our ten species *P. involutus* with a slightly tomentose pileus of a grayish-brown color whose margin is involute, is regarded as edible.†

3. *Species with a deliquescent pileus* (COPRINEAE).

The principal genus of this group is *Coprinus*, some species of which are everywhere common. They may be readily recognized from the fact that the pileus melts to a black inky fluid soon after the maturity of the spores; for this reason they are often known as ink-caps. All of the larger species and some of the smaller ones are edible. The principal edible species are *C. comatus*, three to nine inches high ; easily recognized by its cylindric form and its shaggy pileus which is formed of yellowish-brown scales on a whitish foundation ; young plants have an annulus ; it often grows in great abundance where waste material and rubbish have been dumped. *C. atramentarius* grows in clusters often of many individuals; it has a grayish-brown pileus and very wide crowded lamellae. Like the preceding it is more common in autumn.

C. micaceus is a smaller species and is one of the earliest spring mushrooms. It has a thin pileus and narrow lamellae, the pileus being pale buff or yellowish often with shining particles over the surface and marked with radiating striations. It grows, often in prodigious quantities about the bases of old stumps and is com-

---

*Peck (Bull. N. Y. State Mus. Nat. Hist. 2 : 34–43) gives descriptions of the ten species growing in New York.

†Peck (Bull. N. Y. State Mus. Nat. Hist. 2: 29–33) describes the five species of *Paxillus* growing in New York.

mon even on the streets of our larger cities.* It is commonly eaten raw in salads.

Montagnites is a rare form, known only from Texas and New Mexico in which the pileus is reduced to a simple, disc-like expansion of the end of the stipe. It is interesting only from its relation to certain puff-balls. *Bolbitius* differs from both the other genera in having rusty brown spores. The five American species are mostly small and inconspicuous.

4. *Species with waxy lamellae* (HYGROPHOREAE).

Three genera † make up this group. Of these *Gomphidius* has black spores and decurrent lamellae. *Nyctalis* contains only a single species, peculiar in its parasitic habit; it grows usually on the upper surface of the pileus of large species of *Lactarius*.

By far the greater number of the species belong to *Hygrophorus*, about thirty species being reported from this country. Some of the species are highly colored, *H. miniatus* being a bright red, and *H. psittacinus* having a green pileus and yellow stem.‡

5. *Genera either with a milky juice, or with brittle adnate lamellae and a fleshy stem.* (LACTARIEAE.)

According to Hennings three genera § belong here, but they have usually been considered as forming only two, and it is perhaps best to let them remain so for the present.

Russula is so called from the predominance of species with a red pileus and can be usually recognized by its brittle character, added to its fleshy stem and usually adnate lamellae. A large number of species are described from America,‖ but the limita-

---

* Morgan (Jour. Cincinnati Soc. Nat. Hist. **6**: 173-177) describes thirteen species of *Coprinus* occurring in Ohio; and Massee (Annals of Botany, **10**: 123-184. *pl. 10, 11*. 1896) has given a monograph of the entire genus.

† Hennings (*loc. cit.*) also separates the species of *Hygrophorus*, which have a slimy veil as a distinct genus, *Limacium*.

‡ Peck (Reg. Rep. **23**: 112-114) describes seven of the species then known to occur in New York, but many species have since been reported.

§ *Loc. cit.* 213-221; he separates the genus *Russulina* from *Russula* based on the species with the spores tinged with ochraceous.

‖ Probably thirty species have been described and there are others. Macadam (Jour. Mycol. **5**: 58-64, 135-141) attempted a synopsis of our species, but the work was discontinued after about twenty-five had been described.

tions of the species have not been well defined. Some of the more common species may be distinguished as follows :

A. Species with a red pileus.

R. emetica varying in color from pink to scarlet, with broad mostly continuous lamellae and a strong peppery taste. Its name indicates its effect and it is to be regarded among the poisonous species.

R. lepida, bright or dull red, lamellae somewhat crowded, often forked ; stem white or streaked with pink ; taste mild ; edible.

R. rubra, cinnabar red or vermilion, becoming paler with age ; lamellae crowded, forked and with shorter ones intermixed ; taste acrid ; poisonous.

R. alutacea, from bright to deep red, with buff-yellow lamellae and nutty taste ; edible.

B. Species with a greenish pileus.

R. virescens, an edible species with the grayish-green pileus adorned with small, flocculent scales, and lamellae narrowed as they approach the stem, occasionally with a few short ones intermixed ; edible.

R. heterophylla, greenish or pinkish gray, with milk-white crowded lamellae, often forked, with short ones intermixed ; edible.

Besides these there are others with a pileus of the same colors and some in which the pileus is a clear white. Taste is of special importance in separating the species of this genus ; those with a distinct peppery taste should not be used for food.

The species of *Lactarius* are very easily recognized by their milk, which exudes from the plants at every point when broken or injured. Nearly fifty species are already known from this country.*
Some of the more common species are the following :

A. Species with white milk, which does not change color.

L. piperatus is white and smooth with a depressed or funnel-shaped pileus, with narrow crowded lamellae and a solid stem ;

---

* Peck (Reg. Rep. 38 : 111–133), gives a synopsis of forty of these found in New York with rather full descriptions and notes.

the pileus varies up to four inches wide and the stem is short, varying from a half an inch to two inches long ; taste peppery.

L. volemus is an edible species with a reddish-brown pileus two to five inches across and a stem one to four inches long of nearly the same color as the pileus ; it is commonly gregarious. *L. rufus*, a smaller brighter red species with an acrid peppery taste, is said to be poisonous.

B. *Species with orange-colored milk.*

L. deliciosus, as its name would imply, is an edible species easily recognized by its usually depressed grayish-orange pileus which is marked with brighter zones ; the lamellae have nearly the same color ; when found it is usually gregarious. *L. Chelidonium* has a yellow milk nearly the color of that flowing from celandine.

C. *Species with blue milk.*

L. Indigo is our only species which is characterized by its color, which is almost prussian blue within and a whitish blue without.

6. *Genera with a leathery pileus and grooved or splitting villous lamellae.* (SCHIZOPHYLLEAE.)

Schizophyllum is the only genus of which *S. commune* is everywhere common on standing saplings or on fallen branches, varying from a half an inch to two inches across. A second species has been reported from New Mexico.

7. *Genera with a tough, leathery, thin membranous, or rarely somewhat fleshy, pileus, which revives after drying at the return of moisture.* (MARASMIEAE.)

This tribe is made up chiefly of the genus *Marasmius*, of which we have numerous,* mostly small species. A larger edible species, *M. oreades*, is commonly known as the fairy-ring, because of its habit of growth in circles ; it is very common in pastures. Besides this genus we have a single representative of the tropical genus *Anthracophyllum* in South Carolina, a single species of *Heliomyces* in Alabama, two species of *Nerotus* also southern,

---

* Some sixty species are described from this country. Morgan (Jour. Cincinnati Soc. Nat. Hist. **6** : 189-194), describes seventeen species occurring in Ohio, and Peck (Reg. Rep. **23** : 124-126), describes seven occurring in New York, but many have since been reported.

and several species each of *Panus** and *Lentinus*.† Of the former genus a small phosphorescent species, *P. stypticus*, a half inch or more broad, is very common on rotten wood. Of the latter, *L. lepideus* is common on railroad ties, *L. tigrinus* in wet, open swampy places, and *L. strigosus* on old stumps.

8. *Genera with a fleshy pileus, usually decaying rapidly after the maturity of the spores.* (AGARICEAE.)

This tribe contains all the remaining genera and is by far the largest of the tribes. For convenience the genera may be divided into several smaller groups. Among these the most marked are those possessing a volva in which the young mushroom is completely enveloped and which remains either in the form of a permanent cup at the base of the stem or as a series of floccose separable scales on the surface of the pileus, according as the volva is tough or tender. This group is sometimes known as the VOLVATAE‡ and contains some of the most deadly as well as some of the finest edible species. The species are very common and a single one is probably responsible for most of the fatal cases of mushroom poisoning that have occurred in this country; that it may be more easily detected and avoided we illustrate it (*Pl. 8. f. 1, 2.*). It is known as *Amanita phalloides*; the pileus is smooth, white, greenish or brown, and from three to five inches broad; the lamellae remain white, the base is bulbous and is loosely margined by the volva; it grows in woods, in open places or in the shade of bushes in pastures from July to October. *A. verna* has practically the same poisonous character and is sometimes regarded as the same; it differs mainly in its closer investing volva. Its white lamellae, white spores, and bulbous volvate base will readily enable any one to avoid it.

Amanita muscaria is equally poisonous, but lacks the trim

---

* Hennings unites this genus with *Lentinus*. Forster (Jour. Mycol. 4: 21–26), describes the fourteen species of *Panus* found in this country.

† Twenty-seven American species have been described; Morgan (Jour. Cincinnati Soc. Nat. Hist. 6: 194–196), describes ten of these which grow in Ohio. He has later established the genus *Lentodium* on what had been regarded as a diseased state of *Lentinus tigrinus*, which will probably prove a well-founded genus; it deserves further study.

‡ A useful compilation of descriptions of species in this group has been made by Mr. C. G. Lloyd, of Cincinnati (separately published).

sleekness of *A. phalloides;* it is known as the fly agaric because of its effect in poisoning flies for which it is sometimes used in the country. The pileus is usually a yellow ranging from orange red to pale yellow, the younger specimens being usually brighter ; the lamellae and stem are white and the latter is bulbous at the base. The volva is rather thin and usually leaves only scaly margins at its base, while the greater part is carried up and remains in the form of scattered floccose scales on the pileus. The decoction of this plant is used by the Russians in Siberia for producing hilarious intoxication. We have an allied tho smaller species in *D. Frostiana* which has doubtless often been confused with *A. muscaria;* the latter frequently has a pileus seven or eight inches in diameter while the former rarely reaches more than two.

From these noxious forms we turn to a third member of the group which has long been esteemed as an article of food in Southern Europe, where it is very abundant. In the Italian cities one often sees the peasant women bringing it to the street markets in prodigious quantities. It has been used for food since the time of the old Romans and is known as *Amanita caesarea.* It is equally common in our Southern States and is occasionally found as far north as New York and Massachusetts. It can be easily recognized by its bright orange-red pileus, by the gills, veil (which hangs about the stem like a skirt) and stem, all of which are yellow, and by the persistent cup left by the burst volva. Another species, *A. rubescens,* is regarded as edible, but our advice would be to avoid all volvate species unless, like *A. caesarea,* the marks are absolutely certain.

A second white-spored genus with a volva is *Amanitopsis,* which differs from *Amanita* in lacking a veil. *A. vaginata,* with a thin, usually grayish pileus, which is distinctly striate on the margin, is regarded as edible.*

Other genera of volvate agarics are: *Volvaria,* with salmon colored spores, of which we have seven species, and *Chitonia,* with dark-brown spores, of which we have a single species reported from Nebraska.†

---

* Morgan (Jour. Mycol. 3 : 25-33) gives descriptions of nineteen species of *Amanita* and nine of *Amanitopsis.* Peck (Reg Rep. 33 : 38-49) gives descriptions of fourteen New York species. The species are specially abundant in the South.

† Under the name *Clarkeinda plana.*

A second group of the agarics is known as the ANNULATAE. These lack the volva of the last group but are provided with a veil which becomes an annulus as the pileus expands. Among the white-spored forms we have the genera *Lepiota* and *Armillaria*. The first is well represented with us by nearly thirty species,* mostly characterized by a scaly pileus, free lamellae and a conspicuous, often movable, annulus. The species are often known as parasols; the following larger species are commonly eaten:

L. procera has a brownish or reddish-brown pileus three to five inches across and a very long hollow stem, frequently reaching ten inches and bulbous at the base. *L. rhacodes* resembles it closely but has smaller spores; it is also edible.

L. naucinoides differs from most of the species of the genus in possessing a smooth pileus; it is white throughout with a slightly bulbous base to the stem; the pileus is from two to four inches wide and the stem is from two to four inches long.† It grows commonly in grassy places from August to November and occasionally in cultivated fields.

L. Morgani is a handsome species readily characterized by its green spores; while some people regard it as edible, it has frequently caused sickness, and is to be regarded as a suspicious plant.

Armillaria contains fewer species,‡ but one of them, *A. mellea*, is very common, growing in large masses at the base of old stumps; unlike most of the genus it has a somewhat scaly pileus. The pileus varies from whitish to reddish brown, tho brownish-yellow is probably the most common color; the lamellae tho white at first, finally become stained with reddish brown; the veil varies from cottony to membranous, sometimes disappearing in older plants. While this species is regarded as edible, it is not regarded as specially fine grained or fine flavored.

Among the pink-spored species, *Annularia* has not been found with us; among those with rusty brown spores we have *Pholiota*

* Peck (Reg. Rep. 35: 160-164) has described the eighteen species growing in New York.

† Professor Atkinson figures this as *L. naucina*, a European species with which ours may be identical.

‡ Peck (Reg. Rep. 43: 40-45) gives a synopsis of the six American species with rather full descriptions of the species and their variations found in New York.

with about twenty species, of which some are regarded as edible *;
while among the black-spored types we have a single species of
Anellaria.

It is, however, among the species with dark brown spores that
we have the largest number of annulate species; these are members of the genus *Agaricus*† in its restricted sense which contains
among others the common field mushroom *Agaricus campestris*,
which is also the common mushroom of cultivation. There are
several allied species, almost any of which are edible.‡

Agaricus campestris is easily recognized by a combination of
characters all of which are to be found in the same individual.
Besides the veil which becomes an annulus, and free lamellae, the
plant must have pink lamellae when young changing to dark
brown or nearly black at the maturity of the spores, and a stuffed
stem, *i. e.*, softer in the interior. *A. Rodmani* grows in similar
places, but has the lamellae white at first and narrower than the
thickness of the pileus, and a solid stem. *A. arvensis* and *A. subrufescens* grow in similar localities, but have hollow stems; the
former has an annulus formed of two layers the lower of which is
split into broad yellowish rays; the latter has the annulus floccose
on its lower surface. *A. hemorrhoidarius* grows in woods and may
be recognized by its flesh changing to a dull red when wounded,
while *A. placomyces* and *A. silvaticus* growing in similar situations have white flesh; the former has the pileus covered with
persistent brown scales, while the latter is without scales or has
only evanescent ones. *A. maritimus*, recently described, grows
along the coast of Massachusetts; its flesh shows a pinkish or reddish color when cut or wounded and it has a short solid stem.

Besides *Agaricus* with free lamellae, *Stropharia* has adnate lamellae, but while some species are common, scarcely any are of
economic importance.‡

---

* Morgan (Jour. Cincinnati Soc. Nat. Hist. **6**: 101-104) describes
eleven of these occurring in Ohio.

† Hennings, *loc. cit.*, for some unexplained reason abandons the usual
practice of assigning the generic name *Agaricus* to this group and transfers
to a group of white-spored species.

‡ Some thirteen species are found in America. Peck (Reg. Rep. **36**:
41-49, and **48**: 133-143) describes the seven most common species.

‡ Seven species are recorded from America. Morgan (Jour Cincinnati Soc. Nat. Hist. **6**: 112, 113) describes three of these occurring in

The group VELATAE includes those species in which there is a veil in the young stage, which does not remain as an annulus. In some genera the veil separates entirely from the stem and remains attached to the margin of the pileus like a fringe; in others the veil is like a spider's web and becomes evanescent as the pileus expands. *Hypholoma* is a common example of the first and *Cortinarius* of the second.

Hypholoma has purplish brown spores and includes a number of species representing two very different types.* Of one type *H. incertum* is a common example, with a thin, light-colored pileus, often growing gregariously in lawns. A second group includes mushrooms known as brick-tops,† which grow until late in the season in immense clusters about the bases of old stumps. One of these with a bitter taste, *H. sublateritium*, is reputed poisonous, while *H. perplexum*, with no unpleasant taste, is regarded as edible.

Cortinarius, with arachnoid evanescent veil and rusty brown spores, is one of our largest and most difficult genera.‡ Many of the species appear only late in the season. In studying the species the color of the young plant compared with the old, the viscidity, hygrophaneity, or dryness of the pileus, the taste, and the markings left on the stem by the retreating veil should all be carefully noted. Probably many of the species are edible, and none are reputed poisonous. Among the edible species are:

C. violaceus with a dry dark-violet pileus, two to four inches wide, marked with persistent hairy scales, and a bulbous solid stem three to five inches long.

C. cinnamomeus has a dry fibrillose yellowish or cinnamon-brown pileus, with a slender stuffed or hollow stem, one to three

---

Ohio. Hennings, *l. c.*, joins this genus with the last and gives the combination the name *Psalliota*.

*Eighteen species of *Hypholoma* are reported from America. Morgan (Jour. Cincinnati Soc. Nat. Hist. 6: 113-115), describes the seven specise occurring in Ohio

† Peck (Reg. Rep. 49: 61, 62) gives a synopsis of the six allied species.

‡ Some sixty species have been reported from this country, but no considerable number of them has ever been described in a single English publication. Peck (Reg. Rep. 23: 105-112) describes twenty-one New York species.

inches long. The lamellae are usually yellow, but in the variety *semi-sanguineus* they are blood red.

C. collinitus has a glutinous, yellowish-tawny or yellow pileus two to three inches wide, with a solid stem two to four inches long, which often cracks transversely.

Flammula also belongs to this group and can be easily recognized by its decurrent lamellae and its habit of growing on wood. It has a dozen or more American species. Allied to this genus, but with scarcely apparent veils, are the genera *Naucoria*,* *Inocybe*,† *Tubaria*‡ and *Hebeloma*.§

The remaining genera constitute the EVELATAE, possessing neither volva nor annulus, the pileus not being bound to the stipe with a veil of any kind. In this group Hennings (*loc. cit.*) plays havoc with the genera as usually understood.‖ We treat them here as they have been treated by Fries, although it is more than likely that some changes in nomenclature will be necessary on account of some names having previously been used for other plants. Except for a single scarcely natural sub-division based on the position of the stem, the genera are most simply grouped by the spore characters.

1. *With lateral stems or no stems whatever.*

The oyster mushrooms, belonging to the genus *Pleurotus*, are easily recognized by their white or whitish spores. Some of them

---

* Nineteen specimens of *Naucoria* are reported from the United States; of these Peck (Reg. Rep. **23**: 91-93) describes seven New York species.

† Eight species of *Inocybe* are American; Morgan (Jour. Cincinnati Soc. Nat. Hist. **6**: 104-106) describes all these species since they occur in Ohio.

‡ Two species only are reported from America. *Cf.* Morgan, Jour Cincinnati Soc. Nat Hist. **6**: 109, 110.

§ Eighteen species of *Hebeloma* are American. Peck (Reg. Rep **23**: 95, 96), describes the six New York species then reported.

‖ Hennings, *loc. cit.*, 230-268, combines *Psathyrella* and *Panaeolus* under the genus *Coprinarius*; *Agaricus* (in the usual sense), and *Stropharia* under the genus *Psalliota*; *Crepidotus*, *Pluteolus* and *Galera* under *Derminus*; all the pink-spored genera under *Hyporhodius*; and finally combines *Pleurotus*, *Omphalia*, *Mycena*, *Collybia*, *Clitocybe* and *Tricholoma* under *Agaricus*!!

may be found at all seasons of the year, from late spring to late autumn, often growing in clusters from some old stump or log or even on the standing trunk of some tree; some of the species are very small. Of the edible species (and none are known to be injurious) the following are best known : *

P. sapidus, easily distinguished by the pale lilac-tint to its spores, which is more pronounced where they fall in considerable quantity. The pileus is of a pale-brownish or ash-gray color and measures from three to six or eight inches in width.

P. ostreatus closely resembles the last, but has white spores and frequently has no stem at all. It appears in autumn or late summer.

P. ulmarius has white spores and usually an eccentric stem; it is usually of a firm consistency and frequently cracks are found in the pileus. It commonly grows on elm trees, whence its name, and appears rather late in autumn.

Claudopus† has salmon-colored spores. *C. nidulans* grows in clusters on the underside of logs and is a handsome species, with tomentose pileus and yellowish lamellae.

Crepidotus‡ has rusty-brown spores and is mostly made up of small inconspicuous species growing frequently on much decayed logs.

2. *With a central stem.*

A. *With black spores.*

The forms with black spores are mostly small mushrooms growing on the earth in pastures or occasionally on manure. *Psathyrella*§ has a striate pileus and a cartilaginous stem, and *Panaeolus* ‖ has a fleshy stem and a smoothish pileus.

---

* Some twenty-three species are reported from this country. Peck (Reg. Rep. **39** : 58-67) describes the seventeen species found in New York.

† Five species only are recorded from this country. Peck (Reg. Rep. **39** : 67-69) includes descriptions of all of these.

‡ About fifteen species are found with us. Peck (Reg. Rep. **39** : 69-73) describes the eleven species growing in New York.

§ Some seven species occur in this country. Peck (Reg. Rep. **23** : 102, 103) describes three species growing in New York.

‖ Six or more species grow in this country. Peck (Reg. Rep. **23** : 100-102) describes five of these.

B. With dark brown spores.

Several genera occur here, though most of them are inconspicuous and unimportant from an economic standpoint. *Deconica* may be recognized by its decurrent lamellae; we have only one species. *Psathyra*, with two species, and *Psilocybe*, with eleven, can be distinguished in their early stages by the incurving of the pileus in the latter genus, while it is straight from the first in the former. *Pilosace*, with lamellae free from the stem, contains as yet only a single American representative.

C. With rusty-brown spores.

In this section we have only *Pluteolus* * and *Galera*, † distinguished from each other by the fact that the pileus is viscid and the lamellae free or nearly so in the former genus, and the pileus is hygrophanous or moist and the lamellae attached in the latter.

D. With pink or salmon-colored spores.

Several genera occur in this group; among them *Pluteus* can be readily distinguished by its lamellae being separate from the stem where it joins the pileus; eleven species are known from the United States. *P. cervinus* is one of the commonest species, often appearing early in the season, usually growing from stumps or old logs; the pileus is grayish-brown with whitish lamellae changing to flesh-color as the spores mature; the stem is three to six inches long, nearly equal and solid.

Of the remaining genera *Clitopilus* and *Eccilia* have decurrent lamellae. The latter, with three species, has a cartilaginous stem, while the former, with fourteen American species,‡ has a fleshy stem.

At least two of our species of *Clitopilus* are edible:

C. orcella, sometimes known as the sweetbread mushroom, grows in pastures and open places and has a slightly viscid pileus and a soft context.

---

* Of the eight species of *Pluteolus* five are known from America and they are described by Peck (Reg. Rep. 46 : 58-61).

† *Galera* is a larger genus Peck (Reg Rep. 46 : 61-69) describes our twelve species

‡ All of these species of *Clitopilus* are described by Peck (Reg. Rep. 42 : 39-46).

C. prunulus is more compact, with a dry, pruinate pileus and is slightly larger than the preceding species.

The remaining genera, *Entoloma*, *Nolanea* and *Leptonia*, have adnate or sinuate lamellae; like the species of *Eccilia* they have angular spores. *Entoloma*, with twelve species, has a fleshy stem while that in the other two genera is cartilaginous. *Nolanea*, with seven species, is characterized by its bell-shaped, smooth or papillose pileus and straight margin, while *Leptonia*, with six American species, has a low arched scaly pileus with the margin at first incurved. None of these genera possess species of economic interest and in general it may be said that the pink-spored agarics are the least important of any, though those with rusty brown spores follow closely behind.

E. With white spores.

The white-spored members of the EVELATAE are very numerous and vary greatly in size and habit. Two genera, *Omphalia* * and *Clitocybe*, † have decurrent lamellae and can thus be easily recognized. The former has a cartilaginous stem, while in *Clitocybe* the stem is fleshy and continuous with the substance of the pileus. Some of the species of *Clitocybe* are edible, the most common of which are the following:

C. infundibuliformis, with a funnel-shaped pileus two or three inches across.

C. nebularis and *C. media*, thick and fleshy, the former grayish with close lamellae and the latter brown with lamellae distant.

C. clavipes is top-shaped with a brownish pileus often slightly umbonate.

C. laccata is pinkish, with more or less waxy lamellae. Some are inclined to regard this species and its congeners as forming a distinct genus.

In the same genus is the gorgeous *C. illudens*, a large golden-yellow species, often growing in immense clusters at the base of

---

* *Omphalia* contains twenty-five or more American species. Peck (Reg. Rep. **45** : 32-42) describes twenty-one of these which occur in New York.

† *Clitocybe* contains about forty American species. Morgan (Jour Cincinnati Soc Nat. Hist. **5** : 66-70) describes thirteen species; no complete synopsis of the American species has yet appeared.

old stumps; some of the pilei are nearly eight inches in diameter, the species is unwholesome if not poisonous.

Tricholoma* with adnate or more commonly sinuate lamellae is the only other member of this group with a fleshy stem, consequently a considerable number of species are classed as edible. Among them are:

T. equestre with a viscid yellow pileus two to four inches wide and bright sulphur yellow lamellae; it commonly grows in pine groves. *T. sulfureum* resembles this in color, but the pileus is not viscid; it is regarded as a suspicious species.

T. transmutans has a viscid tawny red or reddish brown pileus about the same size as the last; the stem is two to four inches long and hollow or stuffed.

T. imbricatum resembles the last in size and color, but is distinguished by its pileus not being viscid and by its solid stem; the pileus often appears scaly from the rupturing of the epidermis.

T. personatum is often abundant. The whole mushroom, pileus, lamellae and stem are usually a pale violet lilac color when young, fading somewhat with age.

Hiatula with a delicate plicate pileus is represented by a single species in North Carolina; most of the species are tropical in their distribution. The two remaining genera with sinuate or adnate lamellae are *Mycena*† and *Collybia* ‡; the former with the margin of the pileus straight is reputed to contain edible species, but none are of much prominence. The latter contains several very common species, some of which are said to be injurious; others are commonly regarded edible. Among the commonest species is *C. velutipes*, readily recognized by grow

---

* This genus is largely represented in this country by about fifty species. Peck (Reg. Rep. **44** : 38-64) has given a valuable synopsis with full descriptions of the forty-six species occurring in New York.

† *Mycena* contains about fifty-two American species; no full synopsis has ever been attempted. Peck (Reg. Rep. **23** : 80-84) describes twelve of the species occurring in New York State, but many more have been since reported.

‡ *Collybia* contains fifty-four American species Morgan (Jour. Cincinnati Soc. Nat. Hist. **6** : 173-177) describes thirteen species occurring in Ohio, and Peck (Reg. Rep. **49** : 32-55) describes thirty-four species known from New York.

ing in clusters, often closely packed, from stumps and trunks commonly of elm trees, and by its yellow-brown pileus and its stem, which is darker and velvety at the base.

C. radicata is also very common and may be easily recognized by its distant lamellae and by its twisted stem ending in a root-like base which descends from a bulbous or fusiform enlargement. Both species are classed as edible.

*
* *

The various genera of the order Agaricales as constituted at present show remarkable series of connecting forms combining the various families as now recognized in an interminable network. Not only are the three representative genera of the families Thelephoraceae, Polyporaceae and Agaricaceae connected by the intergrading species of intermediate genera, but each of the three families is similarly connected with other representative genera of all the other families of the order, and outliers connect with various members of the following orders.

A part of the difficulty involved appears to be due to the partial failure to base genera on natural characters, too much use having been made of artificial characters or superficial resemblances. A part of the difficulty however is due to the fact that we have here an extensive evolution of comparatively recent origin in which very many of the steps of the progression are still in existence.

A few of the details of this series of interrelations can be expressed by the following schedule :

BASIDIOMYCETES

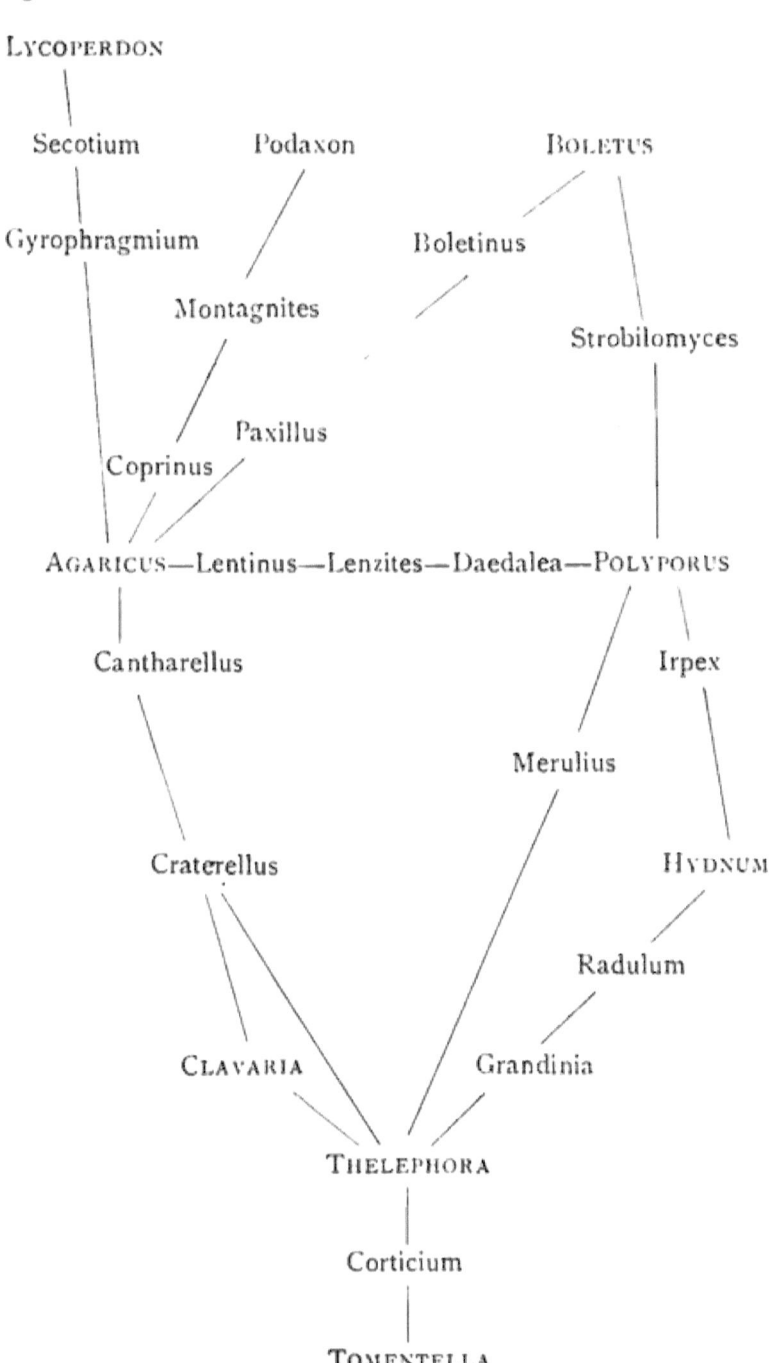

LITERATURE.*

Peck. Reports of the State Botanist of New York, in the Regent's Reports on the State Museum of Natural History, 22-50:
Many of these are difficult to obtain but contain valuable synopses of various genera of Agaricaceae.

Morgan. The Mycologic Flora of the Miami Valley. Jour. Cincinnati Soc. Nat. Hist. 6: 54–81. *Pl. 2–5;* 97–117; 173–199. *Pl. 8, 9.* 1883; 7: 5–10. *Pl.* 1884; 8: 91–110. *Pl. 1;* 168–174. 1885; 9: 1–8. 1886; 10: 7–18; 188–202. 1887; 11: 86–95. 1888. Contains valuable descriptions and notes.

Fries. Hymenomycetes Europaei. 8vo. Upsaliae. 1874.

——— Icones selectae Hymenomycetum nondum delineatorum. 2 vols. folio. Holmiae. 1867–1884. Includes two hundred of the finest plates of mushrooms yet published.

——— Illustrations of British Fungi (Hymenomycetes). 8 vols. 8vo. *Pl. 1–1198.* London. 1881–1891.

Gillet. Les Champignons de la France. Hymenomycetes, 8vo. *Pl. 1–133.* Alençon, 1878; Planches supplementaires. *Pl. 1–384.* Alençon, 1878–1890; Planches supplementaires, suites. *Pl. 1–168.* Alençon, 1890–1895.

To these should be added the older classic works of

Schaeffer. Fungorum Bavariae et Palatinatus icones. 5 vols., 4to. *Pl. 1–330.* Ratisbon, 1780–1800.

Bulliard. Herbier de la France: Histoire des Champignons de la France. 7 vols., 4to. *Pl. 1–612.* Paris, 1784–1812.

Sowerby. Colored Figures of English Fungi or Mushrooms. 3 vols. folio. *Pl. 1–400.* 1797–1803. Supplement. *Pl. 1–39.*

These last contain many illustrations of species belonging to other groups besides the Agaricales.

Saccardo. Sylloge Fungorum, 5: 1–1146 (Agaricaceae); 6: 1–760, 817–928 (Index); 9: 1–257; 11: 1–142.

Hennings. Die natürlichen Pflanzenfamilien, 1¹**: 105–276.

Massee. A Monograph of the Thelephoreae. Jour. Linn. Soc. Botany, 25: 107–155. *Pl. 45–47.* 1889; 27: 95–205. *Pl. 5–7.* 1890.

---

* Much of the descriptive literature has already been cited under the various genera to which it pertains. Some of it is again summarized here together with other more general works; other literature will be found in a later section on edible fungi.

Brefeld. Untersuchungen aus dem Gesammtgebiete der Mykologie, 3 : 1–226. *Pl. 1–11.* 1887 ; 8 : 1–305. *Pl. 1–12.* 1889.

Order 8. PHALLALES.

This order contains a series of most interesting fungi from the standpoint of the botanist, but to most people they present only their disgusting side and are regarded with aversion. The order includes various species known as stink-horns and their unpleasant odor merits for them the name. Many householders have sought in vain for the remains of some dead animal on their premises and have found that the cause of their discomfort was one of these fungi growing unobserved from some buried organic matter somewhere on their lawn. Were the sense of smell wanting or paralyzed, many of the species of this order would be regarded as beautiful objects in nature, but their intolerable odor spoils for most people any natural beauty they possess.

Like many other groups of fleshy fungi they attain their highest development in warmer latitudes. The two families, however, are both represented in this country and may be distinguished as follows :

Receptacle latticed or irregularly branched, sessile or stalked ; gleba enclosed by the receptacle. **Fam 1. Clathraceae.**
Receptacle tubular or cylindric, capitate, with the gleba external.
Fam. 2. Phallaceae.

The Clathraceae are usually rare in the northern states, but in the South, especially in Florida, species of *Clathrus* are common, and their disgusting odor will frequently disclose their presence at long distances. The genera known in America are as follows :

1. Receptacle sessile, latticed, with columnar or obliquely anastomosing bars. CLATHRUS.*
Receptacle ovate or pyriform, nearly sessile ; gleba attached to the inner peridium which ruptures irregularly. PHALLOGASTER.
Receptacle with a tubular stalk. 2.
2. Receptacle latticed, with the meshes of about equal diameter either way.
SIMBLUM.
Receptacle ending in free arms which enclose the gleba when young, but afterwards diverge. ANTHURUS.

---

* The species with columnar bars have sometimes been separated as the genus *Laternea ;* these are the common species of the gulf region.

Of *Clathrus*, three species have been reported from our country; *C. cancellatus* is the only species with oblique lattice work and is as rare as it is beautiful and disagreeable.

Phallogaster saccatus (*Pl. 9. f. 2-4*) is a rare aberrant type showing relations to the Lycoperdales; it probably has a wide distribution but so far has been collected but a few times.

Simblum rubescens was originally described from Long Island and has since been found in Nebraska. (*Pl. 9. f. 1.*) *Anthurus borealis* is our single species from New York and Massachusetts; both are rare species. A species of *Lysurus* was partially described by Ellis* from Texas but nothing further is known of it. The region of the Southwest is likely to give us several additions to this group when properly explored. The ephemeral nature of these plants is such that these species are likely to appear in other localities and will be collected as soon as careful field workers are developed, and it is probable that other species will be found when more extended exploration is made.

The genuine stink-horns belong to the family Phallaceae. Like the members of the preceding family they originate in the form of eggs which rise from mycelial strands connected with a large system of hyphae drawing their nourishment from decaying vegetation. We have seen these interlacing strands extending along an old decaying log for ten feet or more, producing a series of eggs in various stages of development. The genera of this country may be distinguished as follows:

1. Gleba borne on the upper portion of the stipe.   MUTINUS.
   Gleba borne on the surface of a pileus which joins the stipe at the apex.   2
2. With a more or less developed veil hanging like a membrane from the apex of the stipe underneath the pileus.   DICTYOPHORA.
   Veil entirely wanting.   PHALLUS.

Of *Mutinus* we have two or three species fairly well distributed over the eastern and middle states; they have about the diameter of one's finger but are sometimes longer. Two species of *Phallus* †

---

* Bull. Torrey Bot. Club, 7 : 31. 1880.

† By some this generic name has been abandoned; it originated with Micheli in 1729, and was used by Linnaeus with his usual ignorance of affinities to include the morel and the single stink-horn known to him; there is no good reason for substituting the recent name *Ithyphallus* for a name long established, as Fischer and others have done.

occur, *P. impudicus* widely distributed and *P. rubicundus* which is imperfectly known. We have at least two species of *Dictyophora*, *D. Ravenelii* with the stem tapering at each end and with a short veil, and *D. duplicata* with an elegant latticed veil often hanging like a bell-shaped skirt three or four inches wide. The indications are that other species will be found; a second species with a rudimentary veil occurs in the south.

The phalloids have usually been held to be poisonous, but no one with the ordinary powers of smell would think of using them as food. The odor in most cases is useful for the dissemination of the spores since flies are attracted to the gleba and some of the deliquescent mass adheres to their feet and is carried away.

LITERATURE.

Saccardo. Sylloge Fungorum, 7 : 1–27 ; 469, 470 ; 9 : 262–265 ; 11 : 152–156. The Phalloids are treated by E. Fischer.

Morgan. North American Fungi. Jour. Cincinnati Soc. Nat. Hist. 11 : 141–149. *Pl. 3.* 1889.

——— Description of a new Phalloid. Jour. Cincinnati Soc. Nat. Hist. 15 : 171, 172. *Pl. 2.* 1892.

Fischer. Versuch eine systematischen Uebersicht über die bisher bekannten Phalloiden. Jahrb. bot. Gartens zu Berlin, 4 : 1–92. *Pl. 1.* 1886.

——— Untersuchungen zur vergleichenden Entwicklungsgeschichte und Systematik der Phalloideen. Denkschr. Schw. naturf. Gesellsch. 32 :—(1–103). *Pl. 1–6.* 1890.

——— Neue Untersuchungen zur vergleichenden Entwickelungsgeschichte und Systematik der Phalloideen. Denkschr. Schw. naturf. Gesellsch. 33 :—(1–51). *Pl. 1–3.* 1893.

Thaxter. Note on Phallogaster saccatus. Bot. Gaz. 18 : 117–120. *Pl. 9.* 1893.

Burt. A North American Anthurus, its Structure and Development. Mem. Boston Soc. Nat. Hist. 3 : 487–505. *Pl. 49, 50.* 1894.

——— The Phalloideae of the United States. I–III. Bot. Gaz. 22 : 273–292. *Pl. 11, 12;* 379–391. 1896 ; 24 : 73–92. 1897.

Gerard. A new Fungus. Bull. Torrey Bot. Club, 7 : 8–11. *Pl. 1, 2.* 1880.

———— Additions to the U. S. Phalloidei. Bull. Torrey Bot. Club, **7** : 29, 30. 1880.

Order 9. HYMENOGASTRALES.

This order contains a few subterranean genera, most of which are Californian and none of which are common eastward except a single species of *Rhizopogon* which is abundant in some places in the southern states. The plants are tuberous and grow just beneath the surface of the soil, often in sandy places where they are frequently exposed by rain erosion.

The genera of the single family Hymenogastraceae which are found in the United States, may be known as follows :

1. Peridium wanting or obsolete ; spores longitudinally striate. GAUTIERIA.
   Peridium distinct. 2
2. Peridium easily separable. 3
   Peridium not easily separable. 4
3. Spores elliptic or lanceolate, smooth. HYSTERANGIUM.
   Spores globose, echinate at maturity. OCTAVIANA.
4. Peridium surrounded by creeping threads of mycelium. 5.
   Peridium thin, silky villous, with little mycelium or none ; spores ovate or fusiform. HYMENOGASTER.
   Peridium woolly, fleshy, compact ; spores spherical, rough.
   SCLEROGASTER.
5. Peridium thick, somewhat leathery ; spores elliptic, hyaline.
   RHIZOPOGON.
   Peridium thick, tow-like ; spores ovate or elliptic, colored at maturity. MELANOGASTER.

Very few of the species are commonly known and of the dozen species reported from California many if not most are known imperfectly from a single collection. *Gautieria* and *Sclerogaster* have each a single Californian species ; *Octaviana* and *Hymenogaster* have each two representatives in California, and the remaining genera have each three species, all Californian except two species of *Rhizopogon* and a single species of *Melanogaster* from South Carolina. On account of the subterranean habit o the species they are not easy to discover and many other forms are likely to be found. It is probable that there is some relation between the climate and their distribution and they are to be looked for especially on the Pacific Coast and in the Southwest. In addition to the genera included in the synopsis, a species of *Phlyctospora* has been described from Nebraska but little is known either of the species or the genus to which it is said to belong.

LITERATURE.

Saccardo. Sylloge Fungorum, 7 : 154-180, 491, 492 ; 9 : 280, 281 ; 11 : 168-173.

Hesse. Die Hypogaeen Deutschlands. 2 vols., 4to. *IV. 1-22.* Halle, 1891-1894.

Tulasne. Fungi Hypogaei. Folio. *IV. 1-21.* Paris, 1851.

Order 10. LYCOPERDALES.

The ordinary puff-balls form the family Lycoperdaceae* which makes up the present order. In various parts of the country they take the names of smoke-balls or devils' snuff-boxes. The dust of *Bovista pila* is sometimes used in the country for stanching the flow of blood. In their early stages most of the larger species and some of the smaller are used for food. *Calvatia Bovista*, the giant puff-ball, sometimes reaches an enormous size varying from the size of a man's head to that of a half bushel basket, the latter size only rarely reached. Such species develop only where there are buried decaying roots whose substance is widely permeated by the mycelium. As in the preceding order the hyphae are often combined into mycelial strands and often form an extensive net-work, especially in those species which grow in clusters along the bases of very rotten stumps or on logs that have almost fallen to pieces with decay. The young puff-balls appear first as minute balls, and gradually enlarge until the normal size for the species is attained when the fleshy interior which until now appeared white and cheesy, takes on a yellowish or pinkish color and gradually darkens until either the whole mass, or all except a more or less enlarged basal portion, becomes filled with dust-like spores. The spores are usually either yellowish olive-brown or purplish

---

* It is more than likely that when the comparative morphology of this order has been carefully studied, it will be found to contain several family types among its diverse and peculiar forms. Such a division should rest on a more careful comparative research than this group has yet received.

in the various species. The spore-mass is often commingled with thread-like bodies either simple or intricately branched, known as the capillitium; this is commonly attached to the columella or walls, but in some genera it is free. The outer portions often separate into two or three layers and collectively are known as the peridia. The outer peridium in the ordinary puffball separates in small flakes from the inner and falls away at or before the maturity of the spores. In the earth-stars it splits into a series of teeth and becomes flattened or more commonly reflexed. In some forms both layers of the peridium break up into a series of fragments and the spores become scattered either by the wind or more rapidly by the accidental treading of some animal. In others the inner peridium opens by an apical mouth and when compressed belches forth a puff of smoke-like spores. The genus *Catastoma* (*Pl. 9. f. 5–7*) presents a curious anomaly in spore dispersion which was long misunderstood until its true method was brought to light by Mr. Morgan. The outer peridium ruptures in a circumscissile manner about the equator of the puffball its upper portion remaining attached to the apical half of the ball, while the ball itself becomes separated from the lower portion which remains in the ground as a cup. The ball then becoming overturned, opens with a mouth on its original under surface or basal portion, and like the species of some other genera becomes free to be rolled about by the wind, and thus scatters its spores over a wider area. *Myriostoma* (*Pl. 7. f. 7*), a rare but interesting species,* opens by a series of perforations so that the top resembles that of a pepper-box. Some species have a columella formed of a portion of the stem extending up into the spore-chamber; in a few genera this columella extends through to the apex of the peridium; in rare cases the spore cavities are separated by radiating lamellae and thus a series of apparent connecting links unite the puff-balls to the agarics. Thus far a number of peculiar or unique forms have appeared on the Pacific coast and in the semi-arid regions of New Mexico and Arizona, but little is yet known except of the mere existence of these genera. The knowledge of the development, affinities and dis-

---

*This has been reported from only three stations widely separated, in Colorado, Florida and Ontario. *Cf.* Morgan, Am. Nat. **26**: 341, 342, 1892; and Cook, Bot. Gaz. **23**: 43, 44, 1897.

tribution of the species in the genera *Batarrea*, *Polyplocium* and *Podaxon* awaits the careful field work of resident botanists in these regions so fertile in strange fungous productions.

The genera of the Lycoperdaceae can be distinguished as follows :

1. Outer peridium remaining like a volva at the base of the stem. 2.
   Outer peridium splitting into star-like reflexed persistent segments (earth stars). 3.
   Outer peridium becoming gelatinous and later disappearing ; root-like base of interlaced fibers ; lining of inner peridium reddish, appearing at the star like mouth. CALOSTOMA.
   Outer peridium fragile, more or less deciduous, often covered with warts, spines or scales. 5.
2. Inner peridium in the form of a pileus, splitting beneath into thick processes. POLYPLOCIUM.
   Inner peridium circumscissile, the upper part separating like a lid. BATARREA.
3. Inner peridium opening by a single mouth. 4.
   Inner peridium opening by several mouths. MYRIOSTOMA.
4. Columella present ; threads of capillitium simple, tapering to each end. GEASTER.
   Columella wanting ; threads of capillitium long, much branched, interwoven. ASTRAEUS.
5. Peridium with a distinct stalk ; columella if present not extending to the apex. 6.
   Peridium sessile or short stalked with a columella extending to the apex. 7.
   Peridium sessile with a more or less thickened sterile base. 8.
   Peridium sessile without sterile base, spore-bearing throughout ; threads of capillitium free. 10.
6. Peridium closely attached to the stalk, opening by an apical mouth. TYLOSTOMA.
   Peridium readily separating from the stalk, opening irregularly. QUELETIA.
7. Capillitium wanting ; spores borne on more or less lamellar folds. SECOTIUM.*

---

* Curious forms with more distinct lamellae, growing in Lower California, New Mexico and Western Texas, have been variously referred to the genus *Gyrophragmium*, a genus originally founded on an Algerian species. The species are, as yet, too imperfectly known to refer definitely to this or any other genus. Here, again, is an opportunity for local botanists to make a genuine contribution to science.

Capillitium floccose ; columella rigid like a stalk, the peridium opening about its insertion. PODAXON.

Capillitium floccose; peridium splitting laterally; columella floccose. CAULOGLOSSUM.

8. Peridium gradually breaking into fragments from above downwards. CALVATIA.

Peridium opening by a single apical mouth. 9.

9. Capillitium rising from the inner surface of the peridium, long, slender, simple or branched. LYCOPERDON.

Capillitium free, short, several times dichotomously branched. BOVISTELLA.

10. Peridium opening by a basal mouth, the lower part of the outer peridium remaining attached to the soil, the puff-ball becoming overturned and free ; threads of capillitium short, simple or scarcely branched. CATASTOMA.

Peridium opening by an apical mouth ; threads of capillitium short, several times dichotomously branched. BOVISTA.

Peridium opening by the breaking of the upper portion into fragments ; threads of capillitium short with few branches and scattered prickles. MYCENASTRUM.

Calostoma* is represented by three known American species. The earlier stages are unknown, but the orange red peridia can commonly be seen nestling in a translucent covering, reminding one of the yolks of eggs lying in their whites as they are broken into the frying pan. This gelatinous covering disappears later and the root-like base formed of interlacing mycelial fibers supports the peridium whose lining varies from cinnabar red to a brilliant scarlet, the color showing itself at the starlike mouth whose beauty gives the name *Calostoma* to the genus. The species occur from New England to Alabama.

Polyplocium has two species growing in sandy soil in California, but neither is well known. *Batarrea* is a very peculiar genus, one species from New Mexico having a stout stem as thick as a man's thumb, a foot long, with a persistent scaly volva ; the plant forms deep in the ground, its peridium appearing as a cushion-like mass above the surface, its convex portion separating from the flattened basal portion like a lid and exposing the spores and capillitium ; the same or an allied species from California has been

---

* Until recently better known under the later name of *Mitremyces*.

referred to the little known European *B. phalloides*, and still another species has recently been described from Nevada.*

The earth-stars form an exceedingly interesting group of organisms and are represented with us by three well-marked genera. Of these *Astraeus* is the most common, being found everywhere in sandy soil; its outer peridium which in the early stages covers the inner and forms a compact ball, splits into several rigid star-like teeth which spread out when moist and contract when dry so as to again fold over the inner peridium; this peculiar sensitiveness to moisture has given the plant the specific name, *hygrometricus*. When dry it is rolled about by the wind, as it severed its mycelial connections that produced it as soon as its spores were mature, and as it is tumbled along it scatters its spores over a wide extent of territory; when rain or the dewy night overtakes it, it absorbs the moisture and spreads itself out, to take up its endless march as sun and wind again appear to reduce it once more to a ball and set it rolling.

The true earth stars belong to *Geaster* which is represented in this country by nearly twenty species; some of these form stars three inches across while the smallest are often less than an inch; some, like *G. triplex*, show three layers to the peridium; *G. fornicatus* and some others become arched up on their star tips so as to be able to sift out their spores more readily. *Myriostoma*, as noted above, contains a single rare species which has probably been overlooked, and will likely be found to have a wide distribution. *Tylostoma* has a slender stalk usually less than one-fourth of an inch in diameter; a half dozen species occur with us, of which *T. mammosum* is most widely distributed. *Queletia*, another stalked genus, has thicker stems, and opens irregularly instead of apically as in the last named genus; its single species is very rare, having been reported thus far from a single station only.†

Secotium and *Podaxon* are usually rare, though the former is sometimes found abundantly in cultivated fields and pastures; with *Gyrophragmium* it forms an apparent passage to the agarics and for this reason a careful morphological study of these forms is a desideratum. *Cauloglossum* contains one imperfectly known southern species.

---

* *Batarrea attenuata* Peck, Bull. Torrey Bot. Club, 22: 208. 1895.
† Trexlertown, Pennsylvania.

Calvatia contains our largest puff-balls; besides *C. Bovista* already noted, *C. cyathiformis* with purplish spores, and *C. craniiformis* and *C. caelata* with olivaceous spores, are rather widely distributed sometimes growing in horse pastures in early autumn in prodigious quantities; all the forms are edible before the flesh has changed from its fresh white color.

Lycoperdon differs in having an apical mouth, and contains mostly small species ranging from a half inch to two inches in diameter; some have the outer peridium beautifully sculptured into spines and other projections often arranged in patterns; this is the largest genus, containing over thirty American species. *Bovistella*, with one species, *B. Ohiensis*, differs in its free capillitium and is very abundant in the Southern States; its hemispheric sterile bases are frequently long persistent.

Calostoma is represented by three species, and *Bovista* by five; among these *B. plumbea*, an inch or so in diameter, is regarded as a delicacy when young. Finally *Mycenastrum* has a single species distributed from Wisconsin to New Mexico.

LITERATURE.

Saccardo. Sylloge Fungorum, 7: 48-133, 470-488; 9: 266-278; 11: 157-167.

Morgan. North American Fungi. Jour. Cincinnati Soc. Nat. Hist. 12: 8-22. *Pl. 1, 2.* 1889; 163-172. *Pl. 16.* 1890; 13: 5-21. *Pl. 1, 2.* 1891; 14: 141-148. *Pl. 5.* 1892.

A most valuable monograph including descriptions of our species.

Peck. United States Species of Lycoperdon. Trans. Albany Inst. 9: 285-318. 1879. (Separate pp. 34.)

Trelease. The Morels and Puff-balls of Madison. Trans. Wis. Acad. Sci. 7: 105-120. *Pl. 7-9.* 1889.

Massee. A Monograph of the Genus Calostoma Desv. Ann. Bot. 2: 25-45. *Pl. 3.* 1888.

Burnap. Notes on the Genus Calostoma. Bot. Gaz. 23: 180-192. *Pl. 19.* 1897.

Webster. Notes on Calostoma. Rhodora, 1: 30-33. 1899.

Order 11. NIDULARIALES.

This order, with a single family Nidulariaceae, includes the bird's nest fungi, a series of small curious fungi allied to the puff-balls in which the cellular cavities bearing the spores are persistent and remain as sporangioles within the peridium. The species are inconspicuous but rather common. *Cyathus vernicosus*, with ash-colored trumpet-shaped peridia is found in cultivated ground, each peridium containing a few flattened sporangioles within ; *C. striatus*, with brownish peridia, is more commonly found on dried dung. *Crucibulum vulgare*, with yellowish bowls, is found attached to twigs or bits of wood or even to decaying bits of matting in rubbish piles ; this bears a more striking resemblance to a bird's nest after its membranous operculum separates from the mouth of the peridium and exposes its egg-like sporangioles. *Sphaerobolus* has a single spherical sporangiole which is forcibly discharged, probably by the accumulation of the gases of decomposition underneath its viscous sphere. The genera may be distinguished as follows:

1. Peridium double, stellately laciniate, the inner containing a single viscous sporangiole forcibly discharged at maturity.   SPHAEROBOLUS.
   Peridium single.   2.
2. Peridium containing a single sporangiole, setulose at base.   THELEBOLUS
   Several sporangioles in each peridium.   3.
3. Peridium furnished with a deciduous operculum.   4.
   Peridium opening by a lacerate mouth.   NIDULARIA.
4. Peridium tubular-trumpet-shaped of three adnate layers ; spores mixed with filaments.   CYATHUS.
   Peridium bowl-shaped of one cottony layer ; no filaments among the spores.   CRUCIBULUM.

The species are not numerous and those described from this country have never been comparatively studied ; some of the species early described by Schweinitz have never been identified.

LITERATURE.

Saccardo. Sylloge Fungorum, **7** : 28–47 ; **9** : 265, 266 ; **11** : 156, 157.

Tulasne. Recherches sur l'organization et le mode de fructification des champignons de la tribu des Nidulariées, suivies d'un essai monographique. Ann. Sci. Nat. III. **1** : 41–107. *Pl. 3–8.* 1844.

Order 12. SCLERODERMATALES.

The thick-skinned puff-balls of which the common *Scleroderma vulgare* is the type, belong to the Sclerodermataceae. The species are not numerous, but some of them are quite widely distributed and abundant. Our common species grows about old stumps and buried roots and is easily recognized by its rough warty exterior and solid structure which, before maturity, is a bluish lead color within; later it ruptures irregularly to scatter its spores.

Scleroderma flavidum is very abundant southward, often growing in clusters along walks and in partly cultivated ground. *S. Geaster* opens by irregular star-like teeth. The species of *Polysaccum* contain sporangioles so closely packed together that they often become angular; one species in the south frequently becomes as large as the double fists. The American species have never been comparatively studied. The genera reported with us are as follows : *

1. Enclosing sporangioles at maturity. 2.
   No sporangioles at maturity. 3.
2. Wall of peridium single. POLYSACCUM.
   Wall of peridium double, thin. ARACHNION.
3. Peridium distinctly stalked. PHELLORINA.†
   Peridium sessile or with a root-like base. SCLERODERMA.

The last genus is the only one which is in any way common. Two or three species occur in the Northern States and as many more in the South.

LITERATURE.

Saccardo. Sylloge Fungorum, 7 : 133–154, 489–491 ; 9 : 278–280 ; 11 : 167, 168.

This group as well as that of the three preceding orders was elaborated by Dr. J. B. De Toni for Saccardo's Sylloge.

***

*The genus *Cenococcum* hitherto referred here has very uncertain relations and may not belong to this order or even class. Massee (Jour. Mycol. 5 : 184. 1889) has originated a genus *Stella* with a double peridium, the outer star-like, founded on a South Carolina species, but nothing is known of such a plant on this side of the Atlantic.

† This genus is not well known. A single species from California was described by Peck (Reg. Rep. 43: 35).

Edible Fungi.

In the preceding pages allusion has frequently been made to certain fungus species as edible. Some species were used by the Romans as a food and have continued in use to the present day ; all over the continent of Europe many species are commonly seen in the markets and are used by all classes of people. During the season when *Agaricus campestris* appears in abundance in English pastures, special mushroom trains bring the crop to the London markets. In this country less attention has been given to the mushroom as an article of food and except in certain quarters only cultivated forms of *Agaricus campestris* have appeared in the markets, tho in some parts of the central West the morel is commonly eaten as the spring mushroom. With the establishment of mycological clubs throughout the country, the knowledge of the edible species will extend and we may expect in a few years to find them more generally employed as food. Meanwhile the chemists and physiologists are considering their nutritive qualities and at present the indications are that they will discover the fungi much less nutritive than they have been supposed to be.[*] Notwithstanding all this they will continue to be regarded as delicacies, which they really are, and an increased knowledge of their value will very widely extend their use.

With their use as food has come a special literature bearing on the forms desirable to be used for that purpose. The best of this is as follows :

Fries. Sveriges ätliga och giftiga Svampar. Folio, *92 plates*. Stockholm, 1861.

Vittadini. Descrizione dei Funghi mangerecci. 4to, *44 plates*. Milano, 1835.

Gibson. Our edible Toadstools and Mushrooms. 8vo, *38 plates*. New York, 1895.

Cooke. Edible and poisonous Mushrooms. 8vo, *18 plates*. London, 1894.

Peck. Annual Report of the State Botanist of the State of New York. 4to, *44 plates*. Albany, 1896.

---

[*] On this subject *cf.* Mendel. The chemical Composition and nutritive Value of some edible American Fungi. Amer. Jour. Physiology, 1 : 225-238. 1898.

―――― Mushrooms and their Uses. Pamphlet, pp. 35. Cambridge, 1897.

The best simple descriptive work on edible species.

Atkinson. Studies and Illustrations of Mushrooms, I., II. Bull. Cornell Univ. Exp. Station. 138 : 337–366. *f. 87–112.* 1897 ; 168 : 491–516. *f. 83–97.* 1899.

Michael. Führer für Pilzfreunde. 12mo, *55 plates.* Zwickau, 1897.

The best low-priced work with colored plates.

Hay. An elementary Text-book of British Fungi. 8vo. London, 1887.

Specially valuable for its recipes.

CHAPTER IX

FUNGUS ALLIES—THE MYXOMYCETES

(*Slime Moulds, Mycetozoa*)

For many years the group of organisms known under the various names of Myxomycetes, Myxogasters, Mycetozoa, or slime-moulds were associated with the fungi and curiously enough were classed with the puff-balls because of their superficial resemblances. Instead of having relations with this highest order of fungi they stand at the bottom round of plant life, if, indeed, they are plants at all, for some botanists even insist that they are not plants. The zoölogists, however, rarely claim them and these exquisite organisms are likely to be neglected from both sides of the biological household.

A slime-mould is an organism varying in size from a dime to a dinner plate, consisting in the growing stage of a naked mass of yellowish or whitish slimy protoplasm called a plasmodium, possessing a creeping motion, loving darkness rather than light, and living in old rotting logs or stumps, or occasionally on spent tan-bark, or among the rubbish of old chip yards. After a period of feeding on the juices of decay resulting in a growth which extends through a longer or shorter period, it rolls itself up into a ball, or in most cases a series of small balls, forms a thickened wall about its substance and divides into a multitude of microscopic dust-like spores.

To understand its life history more in detail, we may commence with a single spore and follow its course upward until it results in reproducing itself in spores again. The spore on finding itself in a suitable condition of warmth and moisture commences to absorb water, swells up and finally bursts its softened shell and emerges as a mere naked bit of jelly. It then forms from streaming portions of itself delicate cilia, by the lashing of which it manages to wriggle its way about to better feeding grounds than it happened to be left in as a spore. Besides feeding on bacteria it absorbs liquid nutriment rapidly, and soon divides itself into two organisms similar to itself, this division continuing with greater or less rapidity until a large number, usually a swarm, of these bodies are developed, each continuing to feed on its supply of nutriment and increasing to a certain size before division. When a certain stage of this multiplication has been reached, the swarm commences a reverse process, by a union into a mass which ultimately gathers all into its train, adding to itself any stragglers that may be picked up along its line of march. When the individuals are thus absorbed into the plasmodium they lose their cilia and the whole plasmodium progresses from place to place by a seemingly creeping motion which is really the result of a streaming process that occurs within the protoplasm, alternately producing a forward and backward flow. Many species form tongue-like interlacing masses, in which the alternating currents, stronger or weaker, determine the direction of motion of the mass. By cultivating the plasmodium in a moist chamber on wet decaying wood, it is an easy matter to make a demonstration of this streaming motion on glass so that the movement may be made the direct object of microscopic examination. At almost any time during summer and autumn masses of this network of yellowish protoplasm may be found by tearing open soggy decaying logs in the forest.

When the plasmodium has reached a certain size it creeps up to a convenient more or less exposed position on a log or stump or spreads itself over violet leaves, or creeps up the stems of grasses, or clambers over beds of moss or even climbs trees, in order to secure a favorable position in which to produce its spores where they may be more effectively disseminated. In a few cases the whole mass surrounds itself with a more or less tough wall and the interior portion then divides itself into innumerable dust-like spores each one of which is itself surrounded by a thin but impervious wall so as to maintain its contents from completely drying up.* In a greater number of species the plasmodium

---

*Three methods of producing spores result in bodies which receive special names :

1. Stalked or sessile *sporangia* are produced about centers in the plasmodium as the protoplasm assumes an upward direction. These sporangia are of a definite shape for each species and usually quite uniform in size.

breaks up into a large number of separate bodies each of which forms its own covering and develops independently its own spores. In a few cases these spores are produced from the whole mass of the plasmodium within the wall ; in most cases, however, the plasmodium forms both spores and thread-like bodies, known as capillitium, which are often of exquisite designs and are marked in elegant patterns. This serves in various ways the purpose of bursting the wall of the spore-case and more gradually scattering the spores. These spores may in some cases germinate at once or they may remain for a long period before germination. Specimens have been kept for nearly four years without losing their power of germination, and it is quite possible that they would remain a longer time if kept under favorable conditions.

Under certain conditions the plasmodium may take upon itself a more solid form resulting in part from the exclusion of its water and compact itself into a hardened cheesy mass which will remain in a dormant condition for a greater or less time, sometimes for years, before it is awakened into life by the natural recurrence of favorable conditions. In these various ways the forms of slime moulds have maintained themselves as a distinct group and have become widely distributed over the face of the earth.

The time required for the germination of the spores after being placed in water depends on various causes and may occur in as short a space as a half hour or may require nearly a day. The time necessary for a plasmodium to transform itself from a creamy mass of slime to fully formed sporangia will also vary in different species but in the case of *Stemonitis* we have known the whole transformation to take place in the hours between ten at night and five the following morning.

The foot of the stalk is sometimes expanded into a hypothallus and this in turn fuses with similar expansions from other stalks so as to form a continuous membrane.

2. A *plasmodiocarp* elongate, irregular, and often branched or reticulate is formed flat on the substratum from the transformation of the plasmodium.

3. An *æthallium* consisting of a mass formed by the fusion of many sporangia or plasmodiocarps either regularly or more often irregularly. This is sometimes covered with a cortex secreted from the protoplasm.

The organisms belonging to the class Myxomycetes* may be separated into three orders as follows:

Saprophytic; not uniting into definite plasmodia; no common sporangial wall. **Acrasiales.**

Parasitic in living plant-cells, forming a true plasmodium.
Plasmodiophorales.

Saprophytic; forming a true plasmodium from which spores are developed, usually in sporangia. **Myxogastrales.**

The first order contains a few little-known organisms that are found mostly in manure; the American forms have never been specially studied.

The Plasmodiophorales contain a series of parasites mostly attacking the roots of various plants, producing root tubercles. *Frankia alni* quite commonly produces root tubercles on *Alnus*, and *F. ceanothi* is equally common on the roots of *Ceanothus Americanus*. *Plasmodiophora brassicae* causes the so-called club-foot † of cabbage, turnips and other plants of the mustard family, both wild and cultivated; it frequently causes considerable damage.

The true slime-moulds belong to the Myxogastrales. No two writers agree on the division into families, genera, or species, but the following artificial synopsis will probably be useful in enabling a student to recognize the genera found in this country:

1. Spores white, developed on the outside of the plasmodium.
                      CERATIOMYXA.
   Spores developed within sporangia, plasmodiocarps or aethallia.   2.
2. Spores brownish or brownish violet.   3.
   Spores never violet tho usually bright colored.   20.
3. Capillitium present among the spores.   4.
   Capillitium wanting.   PROTODERMIUM.
4. Sporangia with deposits of lime on the outer surface.   5.
   Sporangia without deposits of lime on the outer surface.   13.
5. Sporangia simple.   6.
   Sporangia united in an aethallium.   12.

---

* As stated on p. 19, it would seem best to regard the Mycetozoa as constituting a phylum or primary division of the plant world. The phylum then will contain the single class Myxomycetes.

† This is not to be confused with similar swellings produced in the South and elsewhere on many cultivated plants by nematode worms.

6. Tubules of the capillitium filled with lime throughout. 7.
   Tubules of the capillitium with deposits of lime at the knots, with intervening vacant spaces. 8.
   Tubules of the capillitium without lime. 11.
7. Stipe prolonged into the sporangium as a columella. SCYPHIUM.
   Stipe not entering the sporangium. BADHAMIA.
8. Stipe prolonged as a columella. 9.
   Stipe not entering the sporangium. 10.
9. Sporangium oblong, with a re-entrant apex. PHYSARELLA.
   Sporangium globose, with a convex apex. CYTIDIUM.
10. Sporangia opening with an operculum. CRATERIUM.
    Sporangia opening irregularly. PHYSARUM.*
11. Wall of sporangium with the outer calcareous layer usually compacted into a smooth continuous crust. CHONDRIODERMA.†
    Wall of sporangium bearing minute stellate lime crystals. DIDYMIUM.
    Wall of sporangium with an outer layer of large scales of lime.
    LEPIDODERMA.
12. Aethallia formed of confluent sporangia with columellae, whitish; lime on the surface in the form of stellate crystals. SPUMARIA.
    Aethallia compact, without columellae, yellowish or brownish; lime on the walls in the form of rounded granules. FULIGO.
13. Plasmodium forming simple sporangia. 14.
    Plasmodium forming a large roundish aethallium. AMAUROCHAETE.
    Plasmodium forming elongate irregular plasmodiocarps. 19.
14. Capillitium suspended from the discoid enlargement of the top of the columella, nearly simple. ENERTHENEMA.
    Capillitium growing from a more or less elongate columella. 15.
    Columella wanting; sporangia ovoid, polished. LEOCARPUS.
15. Columella extending nearly to the apex of the sporangium. 18.
    Columella scarcely extending to the center of the globose sporangia. 16.
16. Stipe and columella filled with lime, whitish or yellowish. DIACHAEA.
    Stipe and columella brownish or black. 17.
17. Threads of capillitium several times forked, not forming a net-work.
    CLASTODERMA.
    Threads of capillitium anastomosing to form a net-work.
    LAMPRODERMA.

---

* *Tilmadoche* is separated from this genus by some.

† Morgan adopts the generic name *Diderma* for this, but it is untenable, since it was originally founded by Persoon for a single species, not now regarded as belonging to this genus.

18. Capillitium forming an interior net-work of larger meshes and a superficial net-work of smaller ones. STEMONITIS.
    Capillitium forming only an interior net-work, and reaching the wall of the sporangium with numerous free extremities. COMATRICHA.
19. Plasmodiocarp laterally compressed, splitting into two valves.
    ANGIORIDIUM.
    Plasmodiocarp terete, elongate, opening irregularly. CIENKOWSKIA.
20. Capillitium wanting. 21.
    Capillitium present. 25.
21. Wall of sporangium of uniform texture, or wanting entirely. 22.
    Wall of sporangium with net-like thickenings on the inside forming a sieve-like apparatus by the falling away of the outer layer. 24.
22. Sporangia joined into a flat aethallium, their walls reduced to triangular threads at the six angles of the united sporangia.
    CLATHROPTYCHIUM.
    Sporangia either growing singly, usually irregular in shape, or crowded together but not forming an aethallium. 23.
23. Sporangia grown together nearly their entire length, usually raised on a strongly developed stem-like hypothallus. TUBULINA.
    Sporangia usually of irregular shape, or regular, single or clustered.
    LICEA.
24. Thickenings of parallel ribs connected by much more slender transverse branches. DICTYDIUM.
    Thickenings of net-work thin fibers with enlarged knots. CRIBRARIA.
25. Sporangia or aethallia with columellae. 26.
    Sporangia or aethallia without columellae. 27.
26. Sporangia crowded on a well-developed hypothallus. SIPHOPTYCHIUM.
    Sporangia forming an aethallium with a thick bark. RETICULARIA.
27. Capillitium with irregular or indefinite markings; sporangia sessile with thick brown walls. 28.
    Capillitium with regular and definite thickenings in the form of warts, spines, rings or reticulations. 29.
    Capillitium with regular and definite thickenings in the form of spirals turning to the right. 32.
28. Plasmodium forming roundish or hexagonal sporangia, opening in a circumscissile manner. PERICHAENA.
    Plasmodium forming a more or less elongate, bent and flexuous plasmodiocarp which is irregularly dehiscent. OPHIOTHECA.
29. Sporangia enlarged (0.5–2 cm. in diameter), resembling small puffballs, with a cortex of colored cells. LYCOGALA.
    Sporangia simple, small. 30.

30. Capillitium growing from numerous points in the sporangial wall.
     LACHNOBOLUS.
   Capillitium issuing from the interior of the stipe. 31.
31. Capillitium forming a dense net-work, without free extremities.
     ARCYRIA.
   Capillitium forming a net-work bearing numerous short acute free branches. HETEROTRICHIA.
32. Capillitium arising from the base of the sporangium or the interior of the stipe. 33.
   Capillitium free within the sporangia, forming elaters. 34.
33. Spiral markings of capillitium parallel and conspicuous.
     HEMIARCYRIA.
   Surface of capillitium marked by a system of branching veins which appear at the apices as irregular spirals. CALONEMA.
34. Spiral markings of capillitium parallel and conspicuous. TRICHIA.
   Surface of capillitium marked with irregular spirals. OLIGONEMA.

The slime-moulds are a strictly intermediate group of organisms. In their spore-producing stage they resemble the fungi, but they are not true fungi. In their vegetative or growing stage they resemble certain of the protozoans, but they are not true animals.

It is interesting to note, however, that there are several series of forms of living things connecting the slime-moulds with various other low groups of plants and animals. Connecting links multiply themselves as we continue to investigate the simplest forms of life. We have a series of forms which have been brought to light by an American botanist which seem to connect the slime moulds with certain of the bacteria, and a new order of organisms has been founded as the result of these investigations.* The slime moulds, too, show affinities to some of the Chytridiales parasitic on various diatoms and various filamentous algae. It is claimed by one who has been a diligent student of the slime-moulds, that in certain species he has seen definite attempts at the formation of mycelium, and he regards them as showing some distant affinities to the true moulds. On the animal side we have a series of forms that intergrade almost as perfectly toward distinctive animal types. One, which Haeckel described as *Protomyxa*,

---

* *Cf.* Thaxter, On the Myxobacteriaceae, a new order of Schizomycetes. Bot. Gaz. **17**: 389-406. *Pl. 22-25.* 1892. Also **18**: 29, 30. 1893, and **23**: 395-411. *Pl. 30, 31.* 1897.

passes through a very similar series of stages in its life history from the formation of the ciliated swarm spores and the development of plasmodium to the formation of reproductive bodies. *Protomyxa*, however, lives in water and never leaves it. By some it has been called an aquatic slime-mould and surely shows an affinity with some of the protozoans of which the common amoeba is the type. A wide field is open for searching out affinities among these lowly organisms through cytological study ; at least the present indications lead one to expect considerable light from this kind of investigation.

LITERATURE.

DeBary. Die Mycetozoen (Schleimpilze), ein Beitrag zur Kenntniss der niedersten Organismen. Leipzig. 1864. (See also Comparative Morphology and Biology of the Fungi, Mycetozoa and Bacteria, pp. 421–453).

Rostafinski. Sluzowce (Mycetozoa). 4to, *Pl. 1–13*. Paris. 1875. This work is unfortunately locked up in the Polish language.

Cooke. The Myxomycetes of Great Britain. 12mo, London. 1877.

A translation of the last named work so far as relates to British species.

———— The Myxomycetes of the United States arranged according to the method of Rostafinski. Ann. N Y. Lyc. Nat. Hist. **11**: 378–409. 1877.

Massee. A Monograph of the Myxogasters. 8vo. *Pl. 1–12*. London. 1892.

Lister. A Monograph of the Mycetozoa. 8vo. *Pl. 1–78*. London, 1894.

Morgan. The Myxomycetes of the Miami Valley, Ohio. Jour. Cincinnati Soc. Nat. Hist. **15**: 127–143. *Pl. 3*. 1893 ; **16**: 13–36. *Pl. 1*. 1893 ; **16**: 127–156. *Pl. 11, 12*. 1894 ; **19**: 1–44. *Pl. 1–3*. 1896.

Macbride. The Myxomycetes of Eastern Iowa. Bull. Lab. Nat. Hist. State Univ. Iowa. **2**: 99–162. *Pl. 1–10*. 1892 ; 384–389. *Pl. 11*. 1893.

Schroeter. Die natürlichen Pflanzenfamilien, **1¹**: 1–41. 1889.

Atkinson. The Genus Frankia in the United States. Bull. Torrey Bot. Club, **19** : 171–177. *Pl. 128*. 1892.

SCHIZOMYCETES

* * *

The SCHIZOMYCETES or Bacteria may also be considered as fungus allies (*Cf.* p. 19). Their importance in the world of life ought to be popularly better known, both as agents of contagious diseases, and as agents of fermentation and decay, in which they play a more wholesome part. They are too exclusively studied from a pathogenic standpoint, and their biological and scientific side is unfortunately almost wholly neglected in this country. The special literature on the subject belongs to the specialized science of Bacteriology but for a recent systematic account of families and genera one may consult

Migula. Schizomycetes, in Die natürlichen Pflanzenfamilien 1^{1a}: 1–44. 1896.

CHAPTER X

THE STUDY OF MYCOLOGY IN GENERAL AND ITS STUDY IN AMERICA IN PARTICULAR

While several forms of fungi were known to the ancients and some of them were used as food in Roman times, among the earliest writers to distinguish them by definite characters was a Florentine by the name of Micheli. In 1729 he published a work in folio* which may be said to have laid the foundation of our knowledge of fungus genera as well as that of other cryptogams. This work contained numerous illustrations of fungi and myxomycetes, together with characters of various genera.

In the *Species Plantarum* of 1753 Linnaeus characterized ten genera of fungi, besides *Tremella* which he placed among the algae. These genera are mostly composites, as may be seen from the following list : †

1. AGARICUS, including species in *Cantharellus, Tricholoma, Amanita, Lactarius* and other genera of Agaricaceae, besides *Daedalea* and *Lenzites*. (27 species.)
2. BOLETUS, about equally divided between *Polyporus* and *Boletus* as now understood. (12 species.)
3. HYDNUM, all still included in the genus. (4 species.)
4. PHALLUS, containing *Morchella esculenta* and *Phallus impudicus*. (2 species.)
5. CLATHRUS. (3 species.)
6. ELVELA. (2 species.)
7. PEZIZA, including various Pezizales. (8 species.)
8. CLAVARIA, including also species now in *Cordyceps* and *Xylaria*. (8 species.)

---

* Nova plantarum genera. Florentiae, 1729. Tournefort in 1700 (Institutiones Rei Herbariae), however, had recognized the so-called genera *Fungus, Fungoides, Boletus, Agaricus, Lycoperdon, Coralloides* and *Tubera*.

† Practically the same genera appeared in his Genera Plantarum, 1737.

9. LYCOPERDON, including truffles, myxomycetes and an aecidium, in addition to puff-balls. (9 species.)
10. MUCOR, including *Erysibe*, *Penicillium* and various myxomycetes (11 species.)
11. TREMELLA, placed under the Algae, including an *Auricularia* (?) and a *Gymnosporangium* (?). (2 species.)

These eighty-eight species are each characterized by a single sentence, and were it not for citations of older and better descriptions and figures they would be almost wholly unintelligible. Such was the first summary of the scientific knowledge of fungi less than 150 years ago.

Following this period many species were figured by various authors which, in the absence of type specimens, have become the originals of many of the species of fungi, particularly those of a fleshy character. Among the more prominent of these early writers were Schaeffer (1718-1790) who illustrated the fungus flora of Bavaria and the Palatinate, with nearly four hundred plates (1762-1774), Bulliard (1742-1793) who pictured the fungi of France with even greater display (1784-1795), and Sowerby (1757-1822) who accomplished a similar work for Great Britain (1797-1803). All of these writers gave the most of their attention to the fleshy forms, the agarics coming in for a lion's share, but some of the less conspicuous moulds and parasitic forms were included, particularly in the figures and descriptions of Sowerby. It was Corda (1809-1849), however, who first gave serious attention to the description and delineation of the microscropic characters of the simpler fungi, and his memory needs no further monument than the six folios of *Icones Fungorum* (1837-1854), the last published after his untimely death. The systematic study progressed less rapidly. Persoon (1755-1837) in 1801 published his *Synopsis methodica Fungorum** in which the following genera

---

* Besides this work, Persoon's chief contributions to mycology were:
Tentamen dispositio methodus Fungorum. *Pl. 1-4.* Lipsiae, 1797.
De Fungis clavaeformibus. *Pl. 1-4.* Lipsiae, 1797.
Icones et descriptiones Fungorum minus cognitorum. 4to. *Pl. 1-14.* Lipsiae, 1798.
Icones pictae rariorum Fungorum. 4to. *Pl. 1-24.* Paris, 1803-1806.
Traité sur les Champignons comestibles. *Pl. 1-4.* Paris, 1818.
Mycologia Europaea. 3 vols. *Pl. 1-30.* Erlangen, 1822-1828.

were recognized: *Sphaeria, Stilbospora, Hysterium, Xyloma, Naemospora, Vermicularia, Tubercularia, Sphaerobolus, Thelebolus, Pilobolus, Sclerotium, Tuber, Batarrea, Geastrum, Bovista, Tulostoma, Lycoperdon, Scleroderma, Lycogala, Fuligo, Spumaria, Diderma, Physarum, Trichia, Arcyria, Stemonitis, Cribraria, Licea, Tubulina, Mucor, Onygena, Aecidium, Uredo, Puccinia, Trichoderma, Conoplea, Pyrenium, Cyathus, Clathrus, Phallus, Amanita, Agaricus, Daedalea, Boletus, Sistotrema, Hydnum, Thelephora, Merisma, Clavaria, Geoglossum, Spathularia, Leotia, Helvella, Morchella, Tremella, Peziza, Ascobolus, Helotium, Stilbum, Aegerita, Ascophora, Periconia, Isaria, Botrytis, Monilia, Dematium, Erineum, Racodium, Himantia, Rhizomorpha*, and *Mesenterica*, in all seventy-one genera which represent the second summary of our knowledge of fungi, now nearly one hundred years ago. The real foundation of systematic study, however, dates from the publication of Fries' *Systema Mycologicum* in 1821-32, and on this first extensive summary of our knowledge of fungi the future growth of the system has been built.

In the study of fungi during the present century three names will ever stand prominent, and to each is due the development of a special division of the subject. Elias Fries (1794-1878) laid the foundations for the Basidiomycetes; the brothers Tulasne (Louis René, 1815-1885; Charles, 1816-1884), accomplished a similar work for the Ascomycetes; and Anton De Bary (1831-1888) established the foundations of comparative morphology and biology among the fungi, investigating the questions of their sexuality and relationships to other thallophytes.

Fries was attracted to the study of fungi when a mere child by seeing the magnificent specimens of *Hydnum*, which his native forests of Sweden produced. At the age of twenty-one he published his first mycological paper, and for sixty years he gave his attention especially to the study of the Basidiomycetes. As stated above, his first work of importance was the *Systema* (1821-1832), in which all the known fungi were marshalled in order. To this he published an appendix—*Elenchus Fungorum* (1828), four years before the final volume of the original work appeared—after which he narrowed the range of his studies and in 1836-1838 published his *Epicrisis* in which the known Hymenomycetes of the world were brought up to date. A final revision of the European species appeared in 1874 in his *Hymenomycetes Europaei* the preface of which was dated on his eighty-first birthday. Besides these works bearing on the mycological system he published numerous short papers and several elaborately illustrated folios.*

The brothers Tulasne, after various publications on the tremellines, *Cordyceps*, *Claviceps*, Nidulariaceae, and subterranean fungi, to which citations have already been made under their proper orders, produced their classic work on the Ascomycetes, *Selecta Fungorum Carpologia*, in three folio volumes (1861–1865) in which the fundamental characters of the Ascomycetes are delineated in a sumptuous form that has been the despair of later contributors to the subject.

De Bary after a splendid foundation in his researches among the lower algae took up the special study of the morphology and sexual characters of the lower fungi. His writings, either alone or associated with Woronin, are numerous and widely scattered, but a summary of his conclusions (fortunately translated into English) appears in his Comparative Morphology and Biology of the Fungi, Bacteria and Mycetozoa,† to which the advanced student of mycology must constantly refer as his guide.

In the last generation many important contributions have appeared, two of which deserve special notice. The first is the extensive compilation of Saccardo (1845–), *Sylloge Fungorum*, which involved the transcribing of descriptions of over forty thousand species culled from all languages, their translation into Latin, and their arrangement in an attempted systematic order.‡ While

---

* The two most important of these are cited on p. 131 and p. 144 respectively.

† It is to be regretted that some mycologists have recommended this work to persons who "desired to learn something about fungi." This classic has its proper use, but unless the purpose was to discourage those who wished to learn something of the subject, it should not be recommended to immature students of any age. A beginner cannot well learn Latin with only a copy of Livy in hand, and without a broad preliminary acquaintance with the subject. De Bary is equally unsuited for a novice in mycology.

‡ This work is now becoming rare and expensive. The eleven volumes were published as follows: 1: 1882; 2: 1883 (Pyrenomycetes); 3: 1884 (Sphaeropsideae, Melanconieae); 4: 1886 (Hyphomycetes); 5: 1887 (Agaricaceae); 6: 1888 (Hymenomycetes); 7: 1888 (Gastromycetes, Uredineae, Ustilagineae, Phycomycetes); 8: 1889 (Discomycetes); 9: 1891; 10: 1892; 11: 1895 (supplements).

this work necessarily contains many imperfections and cannot be considered in any sense as a revision, as too many have regarded it, it has made accessible to workers throughout the world the greater part of the technical descriptive literature of the subject and has made possible the recent extensive advances in the definite knowledge of species.*

The other work is that of Brefeld, which is of an entirely different character. His studies have chiefly appeared in the twelve quarto heften, Botanische Untersuchungen über Schimmelpilze, I.-IV. 1872-1881 ; Botanische Untersuchungen über Hefenpilze, V. 1883 ; Untersuchungen aus dem Gesammtgebiete der Mykologie, VI.-XII. 1884-1897, and involve a most elaborate account of the development of fungi from artificial cultures and the study of their comparative morphology. While many new lines of relationship have thus been worked out, Brefeld's sweeping conclusions regarding the origin of the higher fungi from the lower are not being borne out by the work of other investigators, and we are still in the dark in regard to the origin of the higher and more specialized groups.†

A valuable summary of our knowledge of the physiology of the fungi, tho not including the most recent additions, may be found in Zopf, *Die Pilze*.‡ Ludwig's Lehrbuch der niederen Kryptogamen (1892), and Tubeuf and Smith's Diseases of Plants induced by Cryptogamic Parasites (1897), are valuable for students of plant diseases, as are the more elaborate treatises of Soraurer § and Frank.‖ A valuable series of fungi exsiccati parasitic on cultivated plants has been issued by Seymour and Earle under the title Economic Fungi ; eleven fascicles (550 species) have appeared already.

---

* Three annual supplements to the *Sylloge* have appeared as Beiblätter to Hedwigia, 1896, 1897 and 1898, giving a classified index to species of fungi described during the preceding years as follows : 1895—1252 species ; 1896—1313 species ; 1897—1476 species. Surely the end is not yet !

† A somewhat compact summary of Brefeld's system has been made by his assistant, **Von Tafel**, Vergleichende Morphologie der Pilze, Jena, 1892.

‡ In Schenk, Handbuch der Botanik, 4 : 271-781. 1890. Also separately paged.

§ **Soraurer**. Handbuch der Pflanzenkrankheiten. Zweite Auflage. 2 vols. Berlin, 1886.

‖ **Frank**. Die Krankheiten der Pflanzen. Zweite Auflage. 3 vols. Breslau, 1895.

From this very brief survey of the development of the science of mycology in general, we turn to the study of the subject in America in the field and laboratory. This has been a varied history of field workers exploring the varied fungus flora open to them in their immediate vicinity, the attempted correlation of American species with their Old World congeners, and later the extended description of species supposed to be new. This last, until very recently, has fortunately been largely confined to a half dozen workers, so that the types are, for the most part, not very widely separated.

Among the first collectors of fungi in the United States was Louis Bosc (1759–1828), who collected a few species mainly from South Carolina ; he described fifteen species and published figures of fourteen of them in 1811.* It is true that the erratic Rafinesque, three years previously, commenced the publication of a work on the "funguses or mushroom tribes of America," but, fortunately, this was discontinued before much harm had been accomplished.† In 1813 he also published a short paper in Desvaux's Journal de Botanique.

The first extensive study of our fungi was commenced by Lewis David de Schweinitz (1780–1834) who published two principal papers based on his collections near Salem, N. C., and Bethlehem, Pa., where he had served in the capacity of a clergyman of the Lutheran church. His collections were somewhat augmented by contributions from Dr. Torrey, of New York. His collection is now owned by the Philadelphia Academy of Sciences, and though somewhat the worse for the ravages of time, still possesses much of value for the study of his type species.‡ His two principal papers were:

Synopsis fungorum Carolinae superioris secundum observationes Ludovici Davidis de Schweinitz. Edita a D. F. Schwaegrichen. Schr. der naturf. Gesell. Leipzig, 1: 20–131. *Pl. 1, 2.* 1822. and

---

* Gesell. naturf. Freunde Mag. Berlin, 5: 83–89. *Pl. 4–6.*

† Medical Repository, 5: 350–356 ; 356–363. 1808.

‡ In addition to the Schweinitz collection, the Academy at Philadelphia possesses the mycological collection of Dr. George Martin (1827–1886), of West Chester, Pennsylvania ; this is specially full in certain of the groups of Sphaeropsidales, together with some foreign exsiccati, among which is one of the few sets of Rehm's Ascomycetes found in America.

Synopsis Fungorum in America Boreali media degentium. Trans. Am. Phil. Soc. 4: 141–317. *Pl. 19. 1834.* This work notices 3098 species of American Fungi.

The next American collector of note was an Episcopalian clergyman, Moses A. Curtis (1808–1872), who was a native of Massachusetts, but whose mycological work was largely accomplished in North Carolina, where he served his church on Sundays and spent the week days driving over the country in search of fungi and other plants. He divided his extensive collections with Rev. M. J. Berkeley, of England, by whom they were mostly named and published in various papers, mainly in the first four volumes of Grevillea (1872–1876), under their joint names. These species are usually referred to as the "B. & C. species," and have been a source of very much uncertainty to later workers because of the brief imperfect descriptions, rarely extending beyond two short lines of Latin. The difficulty is greatly increased by the fact that the type specimens are in Berkeley's collection at the Royal Botanic Gardens at Kew,* and many of them are small and scrappy.

Curtis received numerous specimens from various other collectors in various parts of the country, who occasionally picked up specimens of the more conspicuous fungi in connection with the field study of the higher plants which mainly absorbed their interest. Among these were: Beaumont and Peters (1810–1888) in Alabama, Lapham (1811–1875) in Wisconsin, Charles Wright (1811–1885) in Connecticut, who afterwards collected extensively the fungi of Cuba, Sprague (1823–   ) in Massachusetts, Olney (1812–1878) in Rhode Island, Michener (1794–1887) in Pennsylvania and Ravenel (1814–1887) in South Carolina. The last named botanist deserves a more extended notice. During the period 1852–1860, he distributed in five fascicles the first series of *exsiccati* or dried specimens of fungi issued in this country under the title *Fungi Caroliniani Exsiccati.* These are much more valuable than the series later issued jointly with M. C. Cooke, under the

---

*Curtis' own collection is preserved in the Cryptogamic laboratory of Harvard University, where the duplicates are somewhat more accessible than the originals at Kew. The Harvard collection also includes the extensive herbarium of Professor W. G. Farlow, containing one of the largest series of European exsiccatae to be found in this country.

title *Fungi Americani Exsiccati*, 1878–1882, whose eight hundred specimens show the effects of what may be called a commercial influence.*

The next prominent American mycologist brings us into the domain of the present generation. For over thirty years Charles H. Peck (1833–) has been state botanist of New York and while by no means neglecting the better known higher flora has devoted himself to the collection, study and description of the fungi of a great state whose fungus flora is consequently better known than that of any other state of the Union. In addition he has described species of fungi from other parts of the country so that his collection contains a large number of type species,† perhaps second in number to any collection of American species. Peck's annual reports are found in the Reports of the Regents of the State Museum, those containing fungi commencing with the 22d issued in 1869 and continuing to the 51st issued in 1898. They contain many descriptions of species and numerous revisions and synopses particularly of the agaric genera represented in New York state. Some of the reports are very rare, and very few complete sets are in existence. Besides these reports he has contributed numerous papers particularly to the Bulletin of the Torrey Botanical Club.

A little later J. B. Ellis (1829–) commenced his work at Newfield, New Jersey, his first paper being published in 1874. Since

---

* Ravenel's own collection, which fortunately contains few types, unfortunately after his death, was sold to the British Museum, so can only be consulted across the water. It is the shame of American botany that so much of its cryptogamic material has been sent across the Atlantic to be named, or to be absorbed in bulk by European museums where it cannot be easily consulted by future American students. It is just as reprehensible to prove a traitor to the botanical interests of one's country as otherwise, and true Americans will see to it that both these practices are discontinued in the future.

† These collections are deposited in the State Capitol at Albany and to the disgrace of New York's politicians be it said that they are not decently cared for. Only the state collection can be consulted at all and that only at great discomfort. The types of collections from outside New York are packed away in attics of New York's stupendous pile at Albany. It is hoped that those in authority will correct this distressing condition before the collections suffer irreparable loss from the ravages of insects.

that time he has published an enormous number of descriptions of new fungi, either alone or associated with Cooke, Martin, Everhart, Kellerman, Holway, Harkness, Dearness, or Bartholomew. His earlier collections were sent to M. C. Cooke in England, and as a result of careless determinations, numerous errors have crept into our conceptions of many species which will need correction by future monographers who come to study Ellis' enormous collection.* This contains by far the largest number of types of any collection of American fungi in existence.

The work of Mr. Ellis, while extending over the entire range of the fungi, has been most extensive among the Sphaeriales and the *fungi imperfecti*. In the former group he has published the only manual that has yet appeared in America attempting to cover descriptions of species of any considerable group. †

In addition to this Mr. Ellis has distributed sets of fungi exsiccati under the title of North American Fungi, of which thirty-six centuries have appeared (3600 specimens). As this was published in an edition of some fifty to sixty sets, some idea can be formed of the enormous labor connected with its issue. A second edition under the title of Fungi Columbiani has also been issued to the number of fourteen centuries.

The work of other botanists who have given attention to mycology, like Bessey, Trelease, Morgan, Farlow, Atkinson, Burrill,

---

* Through the liberality of some of the managers of the New York Botanical Garden, the entire Ellis collection has been secured for this institution, where it is available for students properly prepared for work. In addition to the types, the collection contains extensive series of specimens sent by collectors and correspondents throughout the country, with duplicates of many of the species described by other mycologists, and extensive series sent in exchange by foreign botanists. Added to this is one of the most complete series of foreign exsiccati in this country. Housed as it is in connection with a very complete botanical library and the presence of other extensive collections, it forms the most accessible center of research work in mycology that the country affords. The collections belonging to Columbia University and to the writer are also deposited with those of the New York Botanical Garden.

† **Ellis & Everhart**. The North American Pyrenomycetes, pp. 793. *Pl. 1–41*. 1892. This includes the orders Perisporiales, Hypocreales, Dothideales, Sphaeriales, Aspergillales and Hysteriales as here treated.

Earle, Tracy and others will be noted under the states in which their work has been largely confined. We shall attempt in the following chapter to outline briefly the field work that has been accomplished in the various states of the Union.

CHAPTER XI

THE GEOGRAPHIC DISTRIBUTION OF AMERICAN FUNGI

As will be seen in the summary given below, too little is known of the distribution of our fungi to base even the most simple conclusions. Of the eight thousand or more species reported from America, probably one-half are known from a single collection or from the single station in which they were first discovered. Except in parts of New England, New York, New Jersey, Nebraska, and Kansas, no portions of the country have been systematically studied with a view of determining the extent of their fungus flora. Our knowledge of distribution is therefore based largely on the work of local collectors who have made their observations for the most part in the immediate vicinity of their homes, and we shall try to indicate under the separate states the nature and extent of the collections that have thus been made and the published local lists that have been prepared by the several workers. In order to facilitate reference, the states and territories are arranged alphabetically, and afterwards a brief account of the fungus flora of the remaining portions of North America is added, since our geographic boundaries by no means limit floral or fungal areas. While the study of the fungi of this country commenced in the Carolinas, the flora of the South, outside of those two states and Alabama, is less generally known than that of the North,[*] while the region of the Rocky Mountains, the Great Basin, and the great Southwest is known least of all.

Alabama.

Early collections in this state were made by Beaumont and Peters. Some of the material collected by the latter was distributed by Ravenel in his Fungi Caroliniani exsiccati and Peters' own collection is at the University of Alabama at Tuscaloosa. Later collections were made by Atkinson, B. M. Duggar, Earle,

---

[*] *Cf.* an article by the writer in *Garden and Forest*, 9: 263, 264. 1896.

and the writer in whose personal collections most of the material is contained. Duplicates of a considerable part of all these collections are also to be found in the herbarium of the Alabama Polytechnic Institute at Auburn, whose mycological collection and library is the largest in the entire South. The local lists thus far published are as follows:

Underwood & Earle. A preliminary List of Alabama Fungi. Bull. Ala. Agric. Exp. Sta. 80: 113-283, I-XVII. 1897.

List of 1110 species. Includes also a summary of papers treating of Alabama fungi; these are omitted here, and the following have appeared since this enumeration was published:

Atkinson. Some Fungi from Alabama. Bull. Cornell Univ. 3: 1-50. 1897.

Includes 644 species with numerous new species; a hundred or more are not included in the previous list.

Earle. New or noteworthy Alabama Fungi. Bull. Torrey Bot. Club, 25: 359-368. 1898.

Descriptions of twelve species.

Peck. New Species of Alabama Fungi. Bull. Torrey Bot. Club, 25: 368-372. 1898.

Descriptions of eleven species.

Alaska.

So far as known to us not a single species has been reported from this remote part of our country, nor has a single fungus been collected. From the analogies of high latitudes far less favorably situated, a large and interesting fungus flora may be expected when collecting botanists cease to look upon flowering plants as constituting the whole of the vegetation of a province.

Arkansas.

Except for a few fungi collected by F. L. Harvey when he was connected with the Arkansas Industrial University at Fayetteville, we know almost nothing of the fungus flora of this state. Nothing has been published except a few notes on parasitic forms.

Arizona.

Except a few species of fungi incidentally collected by traveling botanists, almost nothing is known of this region which like the adjoining regions of New Mexico and Texas doubtless possesses an interesting fungus flora correlated with its unusual climatic conditions.

California.

Commencing in the seventies Dr. H. W. Harkness and his assistants collected extensively in the vicinity of San Francisco and in the Sierras. The earlier collections were named by various English mycologists, Cooke, Phillips, Plowright and Vize, so that many of the types are in England; those of Cooke are at Kew. Duplicates of these, together with types of later species, are in the California Academy of Sciences. Many duplicates from Harkness are in the Ellis collection. Later collections have been made by Farlow, Holway, Howe, Blasdale, McClatchie, Purpus, by the writer, and perhaps others who have either resided or traveled in the state. Many of these later collections have found their way in duplicate to the Ellis herbarium. A few, particularly fleshy forms have been described by Peck, some by Morgan, and a few of the Uredinales by Dietel. Probably not less than fifteen hundred species have been reported from the State already, probably not one-third of the number that will eventually be found to make up its flora. The local literature bearing on Californian fungi is correspondingly extensive; the principal papers are the following, but very many descriptions are scattered through other publications with less restricted titles:

Cooke. Californian Fungi. Grevillea, 7: 1–4. 1878; 101, 102. 1879.

List of fifty-seven species, including descriptions.

Cooke & Harkness. Californian Fungi. Grevillea, 9: 6–9. 1880; 81–87. 1881; 12: 83, 84; 92–97. 1884; 13: 16–21. 1884; 111–114. 1885; 14: 8–10. 1885.

Enumeration of numerous species of which nearly two hundred are described as new.

——— Fungi on Eucalyptus. Grevillea, 9: 127–130. 1881.

Notes on forty-two species, of which eighteen are described as new.

——— Fungi of the Pacific Coast. Bull. Cal. Acad. Sci. 1: 13–20. 1884.

List of ninety-one species.

Dietel. Drei neue Uredineen aus Californien. Hedwigia, 32: 29, 30. 1893.

---——— New Californian Uredineae. Erythea, 1: 247–252. 1893; 2: 127–129. 1894.
Descriptions of nineteen species by Dietel and Holway.
———— New North American Uredineae. Erythea, 3: 77–92. 1895.
Descriptions of fourteen species, seven from California.
Ellis & Everhart. New Californian Fungi. Erythea, 1: 145–147. 1893.
Descriptions of six species.
———— New West American Fungi. Erythea, 1: 197–206. 1893.
Descriptions of twenty-nine species partly from California.
Farlow. Notes on some injurious Fungi of California. Bot. Gaz. 10: 346–348. 1885.
Harkness. New Species of Californian Fungi. Bull. Cal. Acad. Sci. 1: 29–47. 1884.
Descriptions of seventy-one new species.
———— Fungi of the Pacific Coast. Bull. Cal. Acad. Sci. 1: 159–176. Pl. 1. 256–271. 1885.
Enumeration of 363 species including descriptions of eleven new species.
Harkness & Moore. Catalogue of the Pacific Coast Fungi. Pp. 46. 1880.
Enumeration of 842 species from California.
Hennings. Fungi Americani-boreales. Hedwigia, 37: 266–276. 1898.
Seventeen new species, including nine from California. Collected by Purpus, and four from Mexico.
Holway. A new Californian Rust. Erythea, 5: 31. 1897.
McClatchie. Flora of Pasadena and Vicinity. Reid's History of Pasadena, 605–649. 1895.
Includes 312 species of fungi.
———— Seedless Plants of Southern California. Proc. So. Cal. Acad. Sci. 1: 337–398. 1897.
Includes 630 species of fungi.
Peck. New Species of Fungi. Bull. Torrey Bot. Club, 22: 198–211. 1895.
Includes among numerous species, eighteen from California.
Phillips. Discomycetes from California. Grevillea, 5: 35, 36. 1876.

List of twenty-two species.

————— Fungi of California and the Sierra Nevada Mountains. Grevillea, 5 : 113–118. *Pl. 87–89.* 1877.
List of sixty-six species with descriptions of fourteen new ones.

————— Fungi of California. Grevillea, 7 : 20–23. 1878.
List of fifty species with descriptions of ten new species.

————— On a new Species of Helvella. Trans. Linn. Soc. II. 1 : 423. *Pl. 48.* 1880.

Phillips & Harkness. Fungi of California. Bull. Cal. Acad. Sci. 1 : 21–25. 1884.
Descriptions of twenty new species.

————— Discomycetes of California. Grevillea, 13 : 22, 23. 1884.
Descriptions of ten new species.

Plowright. Californian Fungi. Grevillea, 5: 74. 1876.
Includes two new species among twenty-four species mentioned.

————— California Sphaeriae. Grevillea, 7 : 71–74. 1878.
Includes ten new species among thirty-nine species mentioned.

Plowright & Harkness. New species of Californian Fungi. Bull. Cal. Acad. Sci. 1 : 26. 1884.
Describes two new species.

Vize. Californian Fungi. Grevillea, 5 : 109–111. 1877 ; 7 : 11–13. 1878.
· Includes twenty-one new species among sixty-six mentioned.

Colorado.

Comparatively little systematic work has been done among the fungi in the Rocky Mountain region generally. Some local work has been done by Cockerell and Baker, more perhaps by visiting botanists, Bessey, Earle, and Tracy, who have spent portions of summers in the mountains ; it is doubtful, however, if more than three or four hundred species have been definitely reported from the state. The papers relating to the local flora are :

Cockerell. Some Fungi of Custer county, Colo. Jour. Mycol. 5: 84, 85. 1889.

Ellis & Everhart. New West American Fungi. Erythea, 1: 197–206. 1893.
Includes several species from Colorado.

Macbride. A new Physarum from Colorado. Bull. Lab. Nat. Hist. State Univ. Iowa **3** : 390. 1893.
Peck. Colorado Fungi. Bot. Gaz. **3** : 34, 35. 1878.
Descriptions of eleven species.
Porter & Coulter. Synopsis of the Flora of Colorado. Misc. Pub. U. S. Geol. Surv. of the Territories, **4** : 1–180. 1874.
Includes list of eight fungi by C. H. Peck, with descriptions of two new species.

Connecticut.

Some material was early collected by Charles Wright and later collections have been made by Thaxter, Setchell, Sturgis, and by the writer, who has spent several summers in the northwestern portion of the state. Except for the publications of the state experiment station, which, in the main, relate to species injurious to cultivated plants, scarcely anything has been published on the local fungus flora.

Delaware.

The flora of this State has been extensively collected by A. Commons, who has in manuscript a long list of Delaware species, and a considerable amount has been issued by Professor Chester in the publications of the experiment station, relating especially to the species injurious to fruit trees. Duplicates of most of Commons' collections are in the Ellis herbarium.

Florida.

Collections in this State have been made by Martin, Ravenel, Calkins, Webber, Swingle, Nash, Rolfs, and the writer. The collections of Webber and Swingle are mainly in the collection of the Division of Vegetable Pathology at Washington.* The others, either in the originals or duplicates, are mainly in the collections of the New York Botanical Garden. Several papers bear directly on the Florida flora and numerous species appear singly in separate publications :
Calkins. Notes on Florida Fungi. Jour. Mycol. **2**: 6, 7, 42, 53, 54, 70, 80, 81, 89–91, 104–106, 126–128. 1886.

---

* This collection should be mentioned among the prominent herbaria of the country. It contains many foreign exsiccati and is particularly full in species of parasitic fungi.

Includes a list of some three hundred species of the larger fungi, collected in the State.

Ellis & Martin. Some new Species of Sphaeriaceous Fungi. Am. Nat. **16**: 809, 810. 1882.
Includes descriptions of four Florida species.

―――― New Species of North American Fungi. Am. Nat. **16**: 1001–1004. 1882 ; **18**: 1147, 1148, 1264. 1884 ; **19**: 76, 77. 1885.
Descriptions of forty-one species, mostly from Florida.

―――― New Florida Fungi. Am. Nat. **17**: 1283–1285. 1883 ; **18**: 69, 70, 188–190. 1884.
Descriptions of twenty-two species.

―――― New Florida Fungi. Jour. Mycol. **1**: 97–101. 1885.
Descriptions of fifteen species, partly from Florida.

Georgia.

Ravenel collected in the vicinity of Augusta and some of these collections appear in his exsiccati ; aside from this we know of no field work having been done in the state ; we note a single paper:

Cooke. North American Fungi. Grevillea, **11**: 106–111. 1883.
Describes twenty-seven species partly from Georgia.

Idaho.

So far as we know, nothing has been done in this state toward making known its fungus flora.

Illinois.

Except among the parasitic forms, comparatively little has been done in this central state. A somewhat thorough exploration for the parasitic forms was organized during the eighties by the State Laboratory of Natural History and extensive collections were made mainly by A. B. Seymour. Under the direction of Professor Burrill, the Uredinales were monographed by Seymour and the Erysibaceae by Earle ; other parts were prepared but unfortunately have never been published, since those that did appear have been exceedingly useful far beyond the state. Professor Earle also made collections in the southern part of the state, and other students and assistants of Professor Burrill, M. B.

Waite, G. P. Clinton and others, have made more or less extensive collections. A considerable amount of this material is preserved in the herbarium of the Illinois State University at Champaign, which is the only collection in the state. Besides a few publications from the experiment station and in horticultural reports on economic species, the publications relating to the state flora are not extensive :

Brendel. Flora Peoriana. Természetrajzi Füzetek, 5 : 299-405. 1882. (Separate, pp. 107.)
Includes list of thirty-nine species of fungi.

Burrill. The Uredineae of Illinois. A list of the species. Proc. Am. Soc. Micros. 8 : 93-102. 1885. (Separate, pp. 10.)
A list of species of Illinois rusts.

Burrill & Earle. Parasitic Fungi of Illinois, Part II. Bull. Illinois State Lab. Nat. Hist. 2 : 387-432. 1887.

Burrill [& Seymour]. Parasitic Fungi of Illinois, Part I. Uredineae. Bull. Illinois State Lab. Nat. Hist. 2 : 141-255. 1885.
Description of the Illinois rusts.

——— New Species of Uredineae. Bot. Gaz. 9 : 187-191. 1884.
Descriptions of thirteen Illinois species.

Pammel. Some Mildews of Illinois. Jour. Mycol. 4 : 36-38. 1888.

Indiana.

Comparatively little, also, is known of the fungus flora of this state. Dr. J. N. Rose made a study of some of the mildews in 1886, and several studies on various parasitic forms were made by H. L. Bolley, who collected in the vicinity of Lafayette. In 1893 a state biological survey was organized and the present writer made two reports while connected with the survey.

The material collected in this period is in part in the herbarium of the writer, and in part in the collection of the Division of Vegetable Pathology at Washington, which had employed E. M. Fisher to collect parasitic fungi in Indiana. Other collections have been made by E. W. Olive, Miss Lillian Snyder, and by Dr. J. C. Arthur, who has had charge of the cryptogamic portion of the survey since 1895. Perhaps 600 species have been reported from the state. Aside from Professor Arthur's collection, which is specially rich in the Uredinales, there is no other of importance in the state. The local literature* is not extensive :

    Arthur. Additions to the Cryptogamic Flora of Indiana. Proc. Indiana Acad. Sci. 1896: 214–216. 1897.

    Rose. Mildews of Indiana. Bot. Gaz. 11 : 60–63. 1886. •
Includes notes on twelve species.

    Olive. A list of the Mycetozoa collected near Crawfordsville, Indiana. Proc. Indiana Acad. Sci. 1897 : 148–150. 1898.
List of forty-three species.

    Snyder. The Uredineae of Tippecanoe county, Ind. Proc. Indiana Acad. Sci. 1896 : 216–224. 1897.

    Underwood. Report of the Botanical Division of the Indiana State Biological Survey. Proc. Indiana Acad. Sci. 1893 : 13–67. 1894.
Includes 482 fungi, two species new, in the list of Cryptogams of the state.

    ——— Report of the Botanical Division of the Indiana State Biological Survey for 1894. Proc. Indiana Acad. Sci. 1894 : 144–156. 1895.
Includes 107 species of fungi additional to preceding list.

    ——— Additions to the published Lists of Indiana Cryptogams. Proc. Indiana Acad. Sci. 1896 : 171, 172. 1897.

Iowa.

This State has been more fortunate than many in having at its State Agricultural College a succession of botanists, who have been able to see that the vegetation of the earth is not exclusively made up of seed-producing plants. The study of the lower plants was commenced by Professor Bessey and his students, and has been continued by his successors, Professors Halsted and Pammel. Professor Macbride, of the State University, has also published considerably on the Myxomycetes and Agaricales, and other collections have been made by J. C. Arthur and A. S. Hitchcock. One of the most enthusiastic collectors is E. W. Holway, of Decorah, whose private collection and library are very exten-

---

    * A bibliography of Indiana botany was prepared by the writer in 1893. *Cf.* Proc. Indiana Acad. Sci. 1893 : 20–30. 1894.

sive. Public collections may be found at the State University at Iowa City, and the Agricultural College at Ames.

Besides the following papers, references to Iowa fungi exist in many scattered general papers, besides publications on economic species published by the experiment station :

Arthur. Descriptions of Iowa Uromyces. Bull. Minnesota Acad. Nat. Sci. 2 : 13–37. 1883.
Descriptions of twelve species.

―――― Preliminary List of Iowa Uredineae. Bull. Iowa Agric. Coll. 151–171. 1885.
List of 134 species.

―――― Memorandum of Iowa Ustilagineae. Bull. Iowa Agric. Coll. 172–174. 1885.
Includes twenty-five species.

Bessey. Bulletin of the Iowa Agricultural College issued by the Department of Botany, 109–174. 1885.
Includes a list of over two hundred fungi, pp. 134–148.

Ellis. New Species of North American Fungi. Am. Nat. 17 : 192–196, 316–319. 1883.
Describes thirty-five fungi partly from Iowa.

Ellis & Halsted. New Iowa Fungi. Jour. Mycol. 4 : 7, 8. 1888.
Descriptions of seven species.

Ellis & Holway. New Fungi from Iowa. Jour. Mycol. 1 : 4–6. 1885.
Descriptions of fifteen species.

―――― New Iowa Fungi. Bull. Lab. Nat. Hist. State Univ Iowa, 3^3 : 41–43. 1895.
Descriptions of four species.

Halsted. Iowa Peronosporeae and a dry Season. Bot. Gaz. 13 : 52–59. 1888.
Notes on twenty-five species.

Macbride. The saprophytic Fungi of Eastern Iowa. The Genus Agaricus, Series I. II. Bull. Lab. Nat. Hist. State Univ. Iowa, 1 : 30–44. 1888.

―――― The saprophytic Fungi of Eastern Iowa. Agaricus. Series, III. IV. V. and the Genus Coprinus. Bull. Lab. Nat. Hist. State Univ. Iowa, 1 : 181–195. 1890.

―――― Common Species of edible Fungi. Bull. Lab. Nat. Hist. State Univ. Iowa, 1 : 196–199. 1890.

———— The Myxomycetes of Eastern Iowa. Bull. Lab. Nat. Hist. State Univ. Iowa, 2 : 99–162. *Pl. 1–10.* 1892 ; 384–389. *Pl. 11.* 1893.

———— The saprophytic Fungi of Eastern Iowa. The Polyporeae. Bull. Lab. Nat. Hist. State Univ. Iowa, 3³ : 1–30. 1895.

Macbride & Allin. The saprophytic Fungi of Eastern Iowa. The Puff-balls. Bull. Lab. Nat. Hist. State Univ. Iowa, 4 : 33–66. 1896.

Macbride & Hitchcock. The Peronosporeae of Iowa. Bull. Lab. Nat. Hist. State Univ. Iowa, 1 : 45–52. 1888.

Pammel. New fungous Diseases of Iowa. Jour. Mycol. 7 : 95–103. 1892.

Kansas.

The earliest reports on the fungus flora of Kansas were published by Professor Cragin of Washburn College, who commenced a biological survey of the state early in the eighties which was soon discontinued. Later collections have been made by Kellerman, Swingle, Carleton, Norton, Hitchcock, and Bartholomew. Many duplicates of these are in the Ellis herbarium. Kellerman & Swingle commenced the distribution of a series of exsiccati, but it was discontinued after two decades had been issued. Probably twelve hundred species or more have been reported from the state. Besides the present list of local literature descriptions of many species are scattered through various papers of Ellis :

Cragin. First Contribution to the Catalogue of the Hymenomycetes and Gasteromycetes of Kansas. Bull. Washburn Coll. Lab. Nat. Hist. 1 : 19–28. 1884 ; 33–42. *Pl. 1.* 1885.

List of 188 species of fungi, with descriptions of eleven new species.

———— Second Contribution to the Catalogue of the Hymenomycetes and Gasteromycetes of Kansas. Bull. Washburn Coll. Lab. Nat. Hist. 1 : 65–67. 1885.

List of twenty species.

———— A Contribution to the Knowledge of the lower Fungi of Kansas. Bull. Washburn Coll. Lab. Nat. Hist. 1 : 67–72. 1885.

List of sixty-nine species.

———— A new Genus and Species of Tremelline Fungus. Bull. Washburn Coll. Lab. Nat. Hist. 1 : 82. 1885.

Describes *Ceracea* from Kansas.

Ellis & Bartholomew. New Kansas Fungi. Erythea, **4**: 1–4, 23–29. 1896.
Descriptions of thirty-six species.
——— New species of Kansas Fungi. Erythea, **4**: 79–83. 1896; **5**: 47–51. 1897.
Descriptions of thirty-three species.
Ellis & Everhart. New West American Fungi. Erythea, **2**: 17–27. 1894.
Descriptions of twenty-nine species, twenty from Kansas.
Ellis & Kellerman. Kansas Fungi. Bull. Torrey Bot. Club, **11**: 114–116, 121–123. 1884.
Descriptions of twenty-six species.
——— New Kansas Fungi. Jour. Mycol. **1**: 2–4. 1885; **2**: 3, 4. 1886.
——— New Species of Kansas Fungi. Jour. Mycol. **5**: 142–144. 1889.
Descriptions of eleven species.
Kellerman. A partial List of the Kansas parasitic Fungi, together with their Host Plants. Bull. Washburn Coll. Lab. Nat. Hist. **1**: 72–81. 1885; also Trans. Kansas Acad. Sci. **9**: 79–86. 1885.
List of 181 species from Kansas.
Kellerman & Swingle. New Species of Kansas Fungi. Jour. Mycol. **4**: 93–95. 1888; **5**: 11–14. *Pl. 1.* 1889.
Descriptions of ten species.
——— New Species of Fungi. Jour. Mycol. **5**: 72–78. 1889.
Descriptions of eight Kansas species.
Norton. A Study of the Kansas Ustilagineae, especially with Regard to their Germination. Trans. St. Louis Acad. Sci. **7**: 229–241. *Pl. 25–29.* 1896.
Notes on thirty-three species.
Peck. New Species of Fungi. Bull. Torrey Bot. Club, **22**: 485–493. 1895.
Includes fifteen species from Kansas.
Smyth. Additions to the Flora of Kansas. Trans. Kansas Acad. Sci. **15**: 60–73. 1898.
Includes list of 169 fungi.
Swingle. A List of the Kansas Species of Peronosporaceae. Trans. Kansas Acad. Sci. **11**: 63–87. 1889.

Walters. Erysipheae of Riley County, Kansas. Trans. Kansas Acad. Sci. **14**: 200–204. *Pl. 1, 2.* 1896.

Notes on sixteen species.

Kentucky.

Very little is known of the fungus flora of this State and probably not more than one hundred species have been reported. Mr. Morgan made a single visit to the southern part of the State and reported a series of interesting forms. Miss S. F. Price has also collected a few species. The local literature, except incidental references to parasitic species issued from the experiment station, is confined to the two following papers :

Ellis. New Fungi. Am. Nat. **16**: 810, 811. 1882.
Describes seven species from Kentucky.

Morgan. Kentucky Fungi. Bot. Gaz. **8**: 156, 157. 1883.
List of twenty-seven species of fungi.

Louisiana.

With the exception of a few species collected by Hale mentioned in Berkeley's Notices of North American Fungi, and a few reported in the early seventies by Featherman, the greater part of the fungi known from this state are those collected by Father Langlois. These are mainly represented by duplicates in the Ellis Herbarium. The swamps of the state ought to furnish a rich field for fungi of the larger sorts, and the diverse floral covering the hosts for an extensive parasitic series.

Ellis & Everhart. New Species of Fungi from various Localities. Jour. Mycol. **2**: 37–42, 87–89, 99–104. 1886.

Descriptions of fifty-four species mostly from Louisiana.

Ellis & Langlois. New Species of Louisiana Fungi. Jour. Mycol. **6**: 35–37. 1890.

Descriptions of eighteen species.

Featherman. Report of Botanical Survey of Southern and Central Louisiana, 1871.

Contains among other plants, a list of twenty-one species of fungi.

———— Third annual Report of the Botanical Survey of Southwest Louisiana, 1872.

Contains among other plants, a list of sixty-eight species of fungi.

Langlois. Catalogue provisoire des Plantes Phanerogames et Cryptogames de la Basse Louisiane. 1887.

A list of some 644 fungi collected in the state.

Maine.

A few fungi were collected early by Rev. Joseph Blake and E. C. Bolles ; most of the later collections have been made by Professor F. L. Harvey and his students. The unexplored portions of Maine largely covered with forests ought to yield a rich harvest when properly explored.

Cooke. Decades of Maine Fungi. Proc. Portland Soc. Nat. Hist. **1**: 179–185. *f. 9.* 1862.

Harvey. Contribution to the Pyrenomycetes of Maine,—I. Bull. Torrey Bot. Club, **23**: 50–58. 1896.

List of 122 species.

———— Contribution to a Knowledge of the Myxomycetes of Maine. Bull. Torrey Bot. Club, **23**: 307–314. 1896 ; **24**: 65–71. 1897 ; **26**: 320–324. 1899.

Notes on 132 species found in the state.

———— Contribution to the Gastromycetes of Maine. Bull. Torrey Bot. Club, **24**: 71–74. 1897.

Notes on twenty-nine species.

Harvey & Knight. Cryptogams collected near Jackman, Maine, August, 1895. Bull. Torrey Bot. Club, **24**: 340–342. 1897.

Includes list of fifty-one fungi.

Maryland.

Except a few notes on economic species and miscellaneous collections made by the Washington botanists in the vicinity of that city, the principal knowledge of the fungi of this State is due to the work of Miss Banning, who has painted a large series and presented them to the New York state collection at Albany.

Banning. Notes on the Fungi of Maryland. Field and Forest, **3**: 42–47, 59–63. 1877.

———— Notes on Fungi. Bot. Gaz. **5**: 5–10, 23. 1880.

———— New Species of Fungi found in Maryland. Bot. Gaz. **6**: 165, 166. 1881.

Characters of four species.

——— Maryland Fungi. Bot. Gaz. **6**: 200-202, 210-213. 1881.

Peck. Fungi of Maryland. Reg. Rep. 44: 64-75. 1891.

List of Maryland species included in Miss Banning's folio of paintings presented to the New York State Museum, and including sixteen new species.

Massachusetts.

Early collections were made in Massachusetts by Sprague, by Hitchcock and by C. L. Andrews, who published a brief account of some fungi collected by him in 1856. Later extensive collections have been made by Professors Farlow and Thaxter, A. B. Seymour, and various instructors and students in botany in Harvard University, by Miss Cora Clarke and others. The flora of the eastern part of the state ought to be well known but comparatively little recent local literature has appeared. The greater part of these collections are naturally at Harvard, but not a few of the duplicates have found their way to the Ellis collection. Many of the fungi of the list published by Tuckerman & Frost belong to the Vermont portion of the territory covered, where Frost did the most of his work. Besides the following, numerous papers on economic species have appeared from the Agricultural Experiment Station at Amherst:

Andrews. Contribution to the Mycology of Massachusetts, Proc. Boston Soc. Nat. Hist. **5**: 321-323. 1856.

List of thirty-six species.

Cobb. A List of Plants growing wild within thirty Miles of Amherst, pp. 51. Northampton, 1887.

Includes essentially the same species as in Tuckerman and Frost's earlier published list.

Farlow. List of Fungi found in the Vicinity of Boston. Bull. Bussey Inst. **1**: 430-439. 1876; **2**: 224-252. 1878.

Hitchcock. Catalogue of the Plants growing without Cultivation in the Vicinity of Amherst College, pp. 64. Amherst, 1829.

A list of 170 species.

——— Catalogues of the Animals and Plants of Massachusetts,

Rep. on Geol. Min. Bot. and Zool. Mass. 525–652. 1835. (Also separate, pp. 142. Amherst, 1835.)
A list of 176 species on pp. 645–649 (125–149).

Sprague. Contributions to New England Mycology, Proc. Boston Soc. Nat. Hist. **5** : 325–331. 1856 ; **6** : 315–321. 1859.
A list of 682 species mainly from Massachusetts.
A supplement was made by Frost ten years later. (See under Vermont.)

Tuckerman and Frost. A Catalogue of Plants growing without Cultivation within thirty Miles of Amherst College, pp. vi, 98. Amherst, 1875.
A list of over 1100 species.

Michigan.

In this state no local publications aside from some notices of a few economic species have been issued, altho collections have been made by Professors Wheeler and Spalding, Mr. G. H. Hicks and others. It is doubtful if there are one hundred species all told recorded in any publication as found in the state.

Minnesota.

During the earlier history of this state Dr. A. E. Johnson made a series of collections of the more conspicuous fungi of the state, but so far as known no specimens exist as sponsors to his list which was published in the first volume of the Bulletin of the State Academy of Science. Besides this a single expedition was made in 1886 to the mining region north of Lake Superior in which Arthur and Holway collected fungi of all sorts. A few other miscellaneous collections have been made by Arthur, Macmillan and others but of the rich flora as a whole scarcely a beginning has been made.

Johnson. The Mycological Flora of Minnesota. Bull. Minnesota Acad. Nat. Sci. **1**: 203–302. 1877; 325–344. 1878.
List of 642 species found in the state.

Arthur & Holway. Report on Botanical Work in Minnesota for the year 1886. Bull. No. 3, Geol. and Nat. Hist. Survey of Minnesota. Pp. 56. 1887.
List of 233 species of fungi, including ten new species.

Mississippi.

All that is known of Mississippi fungi is due to Professors Tracy and Earle who have collected extensively the parasitic species in various parts of the state, notably the region of the gulf. The interesting forms of the higher fungi everywhere abundant have scarcely been touched. The following papers comprise some of the results of the work already accomplished.

Tracy and Earle. New Species of Parasitic Fungi. Bull. Torrey Bot. Club, 22 : 174–179. 1895.

Includes twenty-one Mississippi species.

——— New Species of Fungi from Mississippi. Bull. Torrey Bot. Club, 23 : 205–211. 1896.

Descriptions of twenty species.

——— Mississippi Fungi. Bull. Miss. Agric. Exper. Sta. 34: 1895.

List of 353 species, largely parasitic.

——— Mississippi Fungi. Bull. Miss. Agric. Exper. Sta. 38: 1896.

An additional list of eighty-five species.

Missouri.

Our knowledge of the fungi of this state is due partly to Professor Tracy, who held a position for a time in the State University, to his student, B. T. Galloway, now Chief of the Division of Vegetable Pathology at Washington, and Rev. C. H. Demetrio, who collected in the eighties for Dr. Winter and later for Mr. Ellis. Only a few papers have appeared, the other Missouri species being described in scattered places. Later collections have been made by the workers at the Missouri Botanic Garden at St. Louis.*

Winter. Fungi novi Missourienses. Jour. Mycol. 1 : 121–126. 1885.

Descriptions of twenty-five new species.

Winter & Demetrio. Beiträge zur Pilzflora von Missouri. Hedwigia, 24 : 177–214. 1885. (Separate, pp. 37.)

Galloway. Parasitic Fungi of Missouri. Bot. Gaz. 13 : 213. 1888.

Summary of 362 species collected.

---

* The herbarium of this institution contains the fungus collection of its director, Dr. Trelease, and a considerable number of exsiccati.

Montana.

The earlier exploration of this state was made by F. W. Anderson and Rev. F. D. Kelsey, duplicates of whose collections are in the Ellis collection.* A. B. Seymour made a collecting tour along the Northern Pacific Railroad in 1884, extending as far west as Washington, but a large part of his work was confined to Montana. Later Rydberg, Griffiths, and Williams have collected a considerable number of parasitic species chiefly, but no thoro systematic effort has ever been made to explore the state exhaustively. Compared with many of the older states the local literature is considerable :

Anderson. Supplementary Notes. Jour. Mycol. 5: 82–84. 1889.
List of fifty-three species additional to those of last paper.

———— A preliminary List of the Erysipheae of Montana. Jour. Mycol. 5: 188–194. 1889.
Notes on twelve species.

———— Brief Notes on a few common Fungi of Montana. Jour. Mycol. 5: 30–32. 1889.

Ellis & Anderson. New Species of Montana Fungi. Bot. Gaz. 16: 45–49. *Pl. 7; 85, 86. Pl. 10.* 1891.
Descriptions of twelve species.

Ellis & Everhart. Notes on a species of Coprinus from Montana. The Microscope, 10: 129–131. *Pl. 4.* 1890.

Ellis & Galloway. New Western Fungi. Jour. Mycol. 5: 65–68. 1889.
Descriptions of thirteen species of which twelve are from Montana.

Griffiths. Some Northwestern Erysiphaceae. Bull. Torrey Bot. Club, 26: 138–144. 1899.
Includes numerous references to Montana stations.

Kelsey. Notes on the Fungi of Helena, Mont. Jour. Mycol. 5: 80–82. 1889.
List of seventy-four species.

———— Study of Montana Erysipheae. Bot. Gaz. 13: 285–288. 1889.
Notes on nine species.

---

* The original collection of F. W. Anderson was presented to Columbia University after his death.

Seymour. List of Fungi collected in 1884 along the Northern Pacific Railroad. Proc. Boston Soc. Nat. Hist. **24**: 182-191. 1889.
Includes notes on numerous Montana species.

Nebraska.

In nearly every state the systematic effort in the direction of mycological exploration traces its origin to a one or two men; in no state is this more marked than in Nebraska where the stimulus has come from Professor Bessey. Through him and his students of the Botanical Seminar of the State University of Nebraska a systematic exploration of the entire plant life of the state has been undertaken and the publication of an elaborate state flora has been commenced. The result has been that while in 1884 scarcely anything was known of the mycological flora of the state, to-day it is the best known of any with the exception of New York and New Jersey. Messrs. Webber, Williams, Woods, Shear, J. G. Smith, Bell, Pound, and Clements have been the assistants who have contributed chiefly to this result from the mycological side. The extensive collections are largely in the University of Nebraska at Lincoln where a valuable series of exsiccati may also be found.

Besides considerable literature on economic species, the following local literature may be cited:

Botanical Survey of Nebraska. Reports on Collections made in 1892, 1893, 1894-5, 1895-1896.
Include numerous additions to the fungus flora.

Pound. Notes on the Fungi of economic Interest observed in Lancaster County, Nebraska, during the Summer of 1889. Bull. Neb. Agric. Exp. Sta. **11**: 83-91. 1889.
Notes on seventy-five species.

Saunders. Protophyta-Phycophyta. Flora of Nebraska, Part I. 1894.
Includes the Phycomycetes, pp. 35, 48-53, and 55-60 by Pound and Clements.

Webber. A preliminary Enumeration of the Rusts and Smuts of Nebraska. Bull. Neb. Agric. Exp. Sta. **11**: 37-82. 1889.
Notes on 140 species.

————— Catalogue of the Flora of Nebraska. Rep. Neb. State Board Agric. 1889: 37-162. 1890.

Includes a very full list of the fungi.

———— Appendix to the Catalogue of the Flora of Nebraska. Trans. St. Louis Acad. Sci. 6: 1-47. 1892.
Includes numerous species of fungi.

Nevada.

Except a few species collected by H. W. Harkness in his Californian work we know of no collections of fungi having been made in Nevada. It is therefore practically a *terra incognita*, so far as fungi are concerned.

New Hampshire.

Professor Farlow has spent many summers in the White Mountain region and numerous other botanists have made longer or shorter visits to the same region, but while many species have been published from this region in our scattered literature, and abundance of material exists in many collections, we can refer to only a single paper bearing specifically on the flora of the state.

Farlow. Notes on the Cryptogamic Flora of the White Mountains. Appalachia, 3: 232-251. 1884.

New Jersey.

The study of New Jersey fungi was commenced by Mr. Ellis in the seventies, and most that has been accomplished has been due to his labors. The Philadelphia botanists, W. C. Stevenson, Rex, and others have collected in the vicinity of that city; the New York city botanists, notably Gerard, have collected over the areas adjacent to the Hudson river, and the botanist of the state Experiment Station, Dr. Halsted, has collected largely, especially among parasitic species, in which he was at one time assisted by F. L. Stevens. But the greater part of the exploration has been made by J. B. Ellis, and in addition to the following special local literature, numerous species are scattered through Mr. Ellis' writings, and many species distributed in his North American fungi were collected in the vicinity of his home. Probably two thousand or more species are known from the state.

Britton. Catalogue of Plants found in New Jersey. Report of *State Geologist, 2: 27-642. 1889.

Includes a list of 1705 fungi, prepared by J. B. Ellis and W. R. Gerard.

Cooke & Ellis. New Jersey Fungi. Grevillea, 4 : 178-180. *Pl. 68.* 1876 ; 5 : 30-35. *Pl. 75* ; 49-55. *Pl. 80, 81.* 1876 ; 89-95. 1877 : 6 : 1-18. *Pl. 95, 96.* 1877 ; 81-96. *Pl. 99, 100.* 1878 ; 7 : 4-10 ; 37-42. 1878 ; 8 : 11-16. 1879 ; 9 : 103. 1881.

Includes many species enumerated besides about 300 described as new.

Ellis. New Species of Fungi found at Newfield, N. J. Bull. Torrey Bot. Club, 5 : 45, 46. 1874 ; 6 : 75-77. 1876.

Describes sixteen species.

———— South Jersey Fungi. Bull. Torrey Bot. Club, 6 : 106-109. 1876 ; 133-135. 1877.

Describes twenty-nine species.

———— New Species of North American Fungi. Bull. Torrey Bot. Club, 8 : 64-66, 73-75, 89-91. 1881 ; 9 : 18-20, 73, 74, 98, 99, 111, 112, 133, 134. 1882 ; 10 : 52-54. 1883.

Describes ninety-two species largely from New Jersey.

———— New Species of North American Fungi. Am. Nat. 17 : 192-196, 316-319. 1883.

Describes thirty-five species partly from New Jersey.

———— New North American Fungi. Bull. Torrey Bot. Club, 11 : 17, 18, 41, 42, 73-75. 1884.

Describes twenty-four species partly from New Jersey.

———— New Species of Fungi. Bull. Torrey Bot. Club, 10 : 76, 77, 89, 90, 97, 98, 117, 118. 1883.

Describes twenty-nine species partly from New Jersey.

Halsted. Notes upon Peronosporeae for 1890. Bot. Gaz. 15 : 320-324. 1890.

Remarks on fifteen species.

———— Some Notes upon economic Peronosporeae for 1889 in New Jersey. Jour. Mycol. 5 : 201-203. 1889.

———— Reports of the Botanical Department of the New Jersey Agricultural College Experiment Station for 1892, 1893, 1894, 1895, 1896, 1897, 1898.*

---

* An index by hosts to these and the bulletins issued by the New Jersey Experiment Station bearing on fungous diseases has been prepared by F. L. Stevens. *Cf.* Proc. Columbus [Ohio] Hort. Soc. 11 : 136-147. 1896.

Include numerous notes on fungous parasites preying on useful plants.

Peck. Two new Fungi from New Jersey. Bull. Torrey Bot. Club, **5** : 2, 3. 1874.

────── New Fungi from New Jersey. Bull. Torrey Bot. Club, **6** : 13, 14. 1875.

New Mexico.

A few scattering species were early collected by Dr. H. H. Rusby and Marcus E. Jones which were described by Gerard and Peck. A few parasitic species have been collected by Professors Earle and Tracy, and incidentally by others who have traveled across the state. Professors Wooton and Cockerell have collected a few forms in the vicinity of Las Cruces, including an interesting series of Lycoperdaceae now in our hands for examination. A few scattering descriptions appear in various papers, so that perhaps a hundred species have been definitely reported from the state, which is only the beginning for what promises to be a most interesting region.

We can cite only two papers bearing directly on the local flora:

Gerard. Some Fungi from New Mexico. Bull. Torrey Bot. Club, **8**: 34. 1881.

Peck. New Species of Fungi. Bull. Torrey Bot. Club, **12**: 33–36. *Pl. 49.* 1885.

Describes eleven species partly from New Mexico.

New York.

The Empire state proves itself the one most carefully surveyed of any for its fungus flora. As already stated this is largely due to the labors of its state botanist, Charles H. Peck, who has so long studied its fungi particularly. Numerous other botanists have contributed to this end among whom were Dr. E. C. Howe, W. R. Gerard, Judge G. W. Clinton, Dr. C. E. Fairman, Professors Dudley, Atkinson, Durand, and Duggar of Cornell University, Dr. J. C. Arthur and Mr. F. C. Stewart, at different times botanists at the state experiment station at Geneva, C. L. Shear, who has issued three centuries of New York fungi, O. F. Cook, and the present writer who collected largely in the central portion of the state, and others. Mr. J. B. Ellis, whose home was originally in

St. Lawrence county, also collected extensively in that region, and numerous visitors in the Adirondack and Catskill mountains have made incidental gatherings in those regions. As might be supposed the local literature is extensive; nearly 3000 species are included in Peck's reports alone:

Cooke. New York Fungi. Grevillea, 8: 117-119. 1880.

Day. A catalogue of the native and naturalized Plants of the city of Buffalo and its Vicinity. Bull. Buffalo Soc. Nat. Hist. 4: 65-152. 1882; 153-290. 1883. (Reprint, pp. 215.)

Enumerates 869 species of fungi mostly collected by Judge Clinton.

Fairman. Notes on new or rare Fungi from western New York. Jour. Mycol. 5: 78-80. 1889.

———— Contributions to the Mycology of western New York. Proc. Rochester Acad. Sci. 1: 43-53. *Pl. 3, 4.* 1890.

———— Hymenomyceteae of Orleans County, N. Y. Proc. Rochester Acad. Sci. 2: 154-167. 1893.

Notes on 167 species.

Gerard. New species of Fungi. Bull. Torrey Bot. Club, 4: 47, 48; 64. 1873;—5: 26, 27, 39, 40. 1874;—6: 31, 32. 1875; 77, 78. 1876.

Descriptions of forty-eight new species.

Peck. Reports of State Botanist in Regent's Reports on the Condition of the State Museum of Natural History, 22: 25-106. 1869; 23: 27-135. *Pl. 1-6.* 1872; 24: 41-108. *Pl. 1-4.* 1872; 25: 57-123. *Pl. 1, 2.* 1873; 26: 35-91. 1874; 27: 73-116. *Pl. 1, 2.* 1875; 28: 31-88. *Pl. 1, 2.* 1876; 29: 29-82. *Pl. 1, 2.* 1878; 30: 23-78. *Pl. 1, 2.* 1878; 31: 19-60. 1879; 32: 17-72. 1879;* 33: 11-49. *Pl. 1, 2.* 1883; 34: 24-58. *Pl. 1-4.* 1883; 35: 125-164. 1885; 36: 29-49. 1885; 37: 63-68.† 1885; 38: 77-138. *Pl. 1-3.* 1885; 39: 30-73. *Pl. 1, 2.* 1887; 40: 39-77. 1887; 41: 51-122. 1888; 42: 1-48. *Pl. 1, 2.* 1889; 43: 1-54. *Pl. 1-4.* 1890; 44: 1-75. *Pl. 1-4.* 1891; 45: 1-42. 1893. 46: 1-69. 1893; 47: 1-48. 1894; 48: 1-241, *Pl. A, 1-43.* 1897; 49: 1-69. *Pl. 44-49.* 1897; 50: 75-159. 1898; 51: 267-321. *Pl. A, B, 50-56.* 1898.

---

* Issued only as a public document.

† The principal part of this report was afterwards published in Bull. N. Y. State Mus. 1^2: 1-66. *Pl. 1, 2.* 1887.

———— Synopsis of New York Uncinulae. Trans. Albany Inst. 7: 213-217. *Pl.* 1872.

———— Descriptions of new Species of Fungi. Bull. Buffalo Soc. Nat. Hist. 1: 41-72. 1873.

Descriptions of 142 species.

North Carolina.

As already noted a large amount of the early study of American fungi was undertaken in this state by Schweinitz and Curtis. Some species have been collected more recently by Professor G. F. Atkinson and Gerald McCarthy, but beyond Curtis' long list of nearly twenty-five hundred species, we know comparatively little of the fungus flora of the state. With the diversity of elevation, floral covering and climate possessed by the state careful exploration ought to increase this list greatly.

Curtis. Geological and Natural History Survey of North Carolina. Part 3, Botany. Pp. 156, Raleigh, 1867.

Contains a list of 2392 fungi growing in the state, pp. 83-154.

Schweinitz. Synopsis fungorum Carolinae Superioris. Schriften der naturf. Gesell. Leipzig, 1: 20-131. *Pl. 1, 2.* 1822. (Also separate, pp. 105.)

Enumeration of 1373 species, many of them new.

———— Synopsis fungorum in America Boreali media degentium. Trans. Am. Phil. Soc. 4: 141-316. *Pl. 19.* 1834.

Includes 3098 species of fungi, many from North Carolina, including many described as new.

North Dakota.

Beyond some work chiefly on species of economic importance by Professor Bolley at the state experiment station at Fargo, comparatively little field work has been done in the state. A. B. Seymour collected a number of species on his journey in 1884 and his paper which belongs equally to Montana is the only one we can cite as relating specially to the local flora :

Seymour. List of Fungi collected in 1884 along the Northern Pacific Railroad. Proc. Boston Soc. Nat. Hist. 24: 182-191. 1889.

Includes notes on numerous North Dakota species.

Ohio.

Thomas G. Lea was one of the first to study the fungi of Ohio commencing in the early thirties, and his list published in 1849 laid the foundation for subsequent work carried on by Mr. Morgan and others. Many of Lea's specimens were sent direct to Berkeley and the types of several new species together with other material are preserved at Kew. Sullivant also collected some material in the central portion of the state and some of this was described by Montagne, whose types are to be sought at Paris. Later Mr. Morgan has given extensive attention to the various groups of the higher fungi as well as to the Myxomycetes, and his publications include many new species as well as careful descriptions of hundreds of old ones most of which have been published in the Journal of the Cincinnati Society of Natural History. In the case of the more fleshy fungi it is Mr. Morgan's practice to preserve only a painting. His collection contains many hundreds of these all made by the hand of Mrs. Morgan who has aided in bringing together a most valuable collection. Mr. Morgan's work, which has covered nearly every group except parasitic forms, has been ably supplemented on the parasitic side by Professor Kellerman who has collected more extensively in these groups. Other collections, particularly of parasitic species, have been made in the state by A. D. Selby, F. L. Stevens, Miss Detmers, J. F. James and others. More recently Mr. C. G. Lloyd, of Cincinnati, has established a very extensive private collection and library and has devoted a large amount of time to the production of photogravures and photographs of the larger fungi, and to the publication of notes on various species and compiled descriptions which he has distributed widely and with a generous hand. Besides these publications which contain as much of general as local value, and various publications bearing on economic species emanating from the experiment station, the following papers relate to Ohio fungi directly:

Detmers. A preliminary List of the Rusts of Ohio. Bull. Ohio Agric. Exp. Sta., 5: 133–140. 1892.

———— Additions to the preliminary List of the Uredineae of Ohio. Bull. Ohio Agric. Exp Sta (Tech. Ser.), 1: 171–180. 1893.

Ellis & Kellerman. New Species of North American Fungi. Am. Nat. 17: 1164-1166. 1883.
Describes fourteen species chiefly from Ohio.

James. Catalogue of the Flowering Plants, Ferns and Fungi growing in the Vicinity of Cincinnati. Jour. Cincinnati Soc. Nat. Hist. 2: 42-68. 1879. (Separate, pp. 27.)
List of 319 fungi from Lea's earlier catalogue.

Kellerman & Werner. Catalogue of Ohio Plants. Geology of Ohio, 7^2: 56-406. 1895.
Includes a list of some 1080 species known from the state. Includes also a bibliography of Ohio botany.

Lea. Catalogue of Plants, native and naturalized, collected in the Vicinity of Cincinnati During the Years 1834-1844. Pp. 17. Philadelphia, 1849.
List of 319 species of fungi with descriptions of fifty-three new species by Berkeley.

Morgan. The Mycologic Flora of the Miami Valley. O. Jour. Cincinnati Soc. Nat. Hist. 6: 54-81. *Pl. 2, 5;* 97-117; 173-199. *Pl. 8, 9.* 1883; — 7: 5-10. *Pl. 1.* 1884: — 8: 91-111. *Pl. 1;* 168-174. 1885; 9: 1-8. 1886; — 10: 7-18; 188-202. 1887; — 11: 86-95. 1888.
Descriptions of the hymenomycetous fungi of the region.

——— The Myxomycetes of the Miami Valley. Ohio Jour. Cincinnati Soc. Nat. Hist. 15: 127-143. *Pl. 3.* 1893; 16: 13-36. *Pl. 1.* 1893; 127-156. *Pl. 11, 12.* 1894; 19: 1-44. *Pl. 1-3.* 1896.

Selby. The Ohio Erysipheae. Bull. Ohio Agric. Exp. Sta. (Tech. Ser.), 1: 213-224. 1893.

Stevens. Parasitic Fungi on Ohio Weeds. Jour. Columbus Hort. Soc. 11: 120-126. 1896. (Separate, pp. 1-7.)

Oklahoma.

Scarcely anything is known of the fungi of this territory, which is doubtless closely allied to that of the adjoining state of Kansas, which is known to be rich particularly in parasitic species.

Oregon.

Various incidental collections have been made in this state by W. M. Carpenter, W. C. Cusick and Professor F. E. Lloyd, most of which are represented by duplicates in the Ellis collection, but, except for a few station bulletins on economic species, there is no local literature bearing on the fungi of the state.

Pennsylvania.

Mühlenberg included a list of over two hundred species as early as 1813, and later Schweinitz collected extensively near Bethlehem and reported many fungi therefrom in his later work. Michener also collected and exchanged with Curtis and many of his specimens thus found their way to the Berkeley collection now at Kew. Later collections have been made by Dr. Martin, B. M. Everhart, Haines, Gentry and Stevenson, most of which are represented by duplicates in the Ellis herbarium. Dr. George A. Rex, of Philadelphia, also collected many fungi and gave particular attention to the myxomycetes, on which he published several valuable papers. His voluminous correspondence with Mr. Morgan contains a mine of the most valuable notes on this group, often accompanied by exquisite drawings. His untimely death was a great loss to American mycology. Harold Wingate also collected, in this group, largely about Philadelphia. While descriptions of many species from Pennsylvania may be found in widely scattered publications, the number of papers relating primarily or mainly to Pennsylvania species is very small:

Ellis. New Species of Fungi. Bull. Torrey Bot. Club, 10: 76, 77, 89, 90, 97, 98, 117, 118. 1883.
Describes twenty-nine species partly from Pennsylvania.

Mühlenberg. Catalogus Plantarum Americae Septentrionales. Pp. 112. Lancaster, Pa., 1813. 2d ed., 1818.
List of 201 fungi mostly from Pennsylvania.

Schweinitz. Synopsis Fungorum in Americae Boreali media digentium. Trans. Am. Phil. Soc. 4: 141–316. *Pl. 19*. 1834.
3098 species included, partly from Pennsylvania.

Rhode Island.

The fungi of Rhode Island were early collected by J. L. Bennett and S. T. Olney and were largely sent to Curtis and Berkeley for determination; a summary of some 580 species has been published which includes practically all that has been done as yet in the state. The list could easily be trebled by careful systematic work.

Bennett. Plants of Rhode Island. pp. 128. Providence, 1888.

South Carolina.

As before noted it was in this state that the earliest work on American fungi was done. Later work by Ravenel has also been noted; besides the following papers, very many of the fungi of Ravenel's two series of exsiccati were collected from the vicinity of Aiken:

Cooke. North American Fungi. Grevillea, 11: 106–111. 1883. Describes twenty-seven species partly from South Carolina.

Ravenel. Contributions to the Cryptogamic Botany of South Carolina. Charleston Med. Jour. and Rev. 6: 190–199. 1851. List of 169 hymenomycetous Fungi.

———— A list of the more common native plants of South Carolina. South Carolina, Resources and Population, etc. 312–359. 1883.

A list of thirty-five species of fungi, pp. 353–356.

Thümen. Fungorum Americanorum, triginta species novae. Flora, 61: 177–184. 1878.

Twenty-five new species from South Carolina.

South Dakota.

Extensive collections have been made by Professor T. A. Williams, David Griffiths, and various students of the Agricultural College, notably among parasitic species but comparatively little publication has been made regarding local distribution:

Williams. Notes on parasitic Fungi observed at Brookings during the Summer of 1891. Bull. S. D. Exper. Sta. 29: 29–52. 1891.

Notes on ninety-two species.

Griffiths. Some Northwestern Erysiphaceae. Bull. Torrey Bot. Club, 26: 138–144. 1899.

Tennessee.

We know practically nothing of the fungus flora of this state, since neither mycologist nor collector has ever interested himself in fungi. It is to be doubted if references to a score of species occurring in the state could be found in the whole range of our literature.

Texas.

When area is considered, probably this state presents the best example of a region practically unknown to the mycologist. Ravenel collected a number of fungi in the state during a single tour of exploration ; some of these were described by Cooke, but many more, of which duplicates are in the Ellis collection, have never been reported. In addition to this the report of a single series of parasitic species from the experiment station constitutes our entire knowledge of the fungi of one of the most interesting regions in the entire country.

Cooke. The Fungi of Texas. Jour. Linn. Soc. 17: 141–144. 1880.

Descriptions of twenty-five species.

—————— The Fungi of Texas. Ann. N. Y. Acad. Sci. 1: 177 187. 1878.

A list of 149 species.

Jennings. Some parasitic Fungi of Texas. Bull. Texas Agric. Exper. Sta. 9: 23–29. 1890.

List of ninety-five species from the state.

Ravenel. Report on the Fungi of Texas. Rep. Comm. Agric. on Disease of Cattle in United States, 171–174. 1871.

General account of fungi of the state.

Utah.

Sporadic collections chiefly of parasitic species have been made by Marcus E. Jones and S. J. Harkness most of which have been sent to Mr. Ellis. A few have recently been described by European botanists. The field, however, is practically a virgin one like the greater portion of the Rocky Mountain area. The following papers contain descriptions of Utah species, and a few others are scattered through our literature :

Ellis. New species of North American Fungi. Bull. Torrey Bot. Club, 8: 64–66, 73–75, 89–91. 1881; 9: 18–20, 73, 74, 98, 99, 111, 112, 133, 134. 1882; 10: 52–54. 1883.

Includes ten species from Utah among many others.

——— New Ascomycetous Fungi. Bull. Torrey Bot. Club, 8: 123–125. 1881.
Descriptions of thirteen species.

Peck. New species of Fungi. Bull. Torrey Bot. Club, 11: 49, 50. 1884.
Descriptions of nine species partly from Utah.

Vermont.

Much of the early work on Vermont cryptogams was done by C. C. Frost (1805–1880), "the Brattleboro shoemaker," and in 1875 he joined Professor Hitchcock of Amherst in the publication of a list of plants growing within thirty miles of Amherst College, the limit being thus taken to include Frost's tramping grounds in the southeast corner of Vermont. Many of the plants of that list, particularly the cryptogams, belong to Vermont instead of Massachusetts, tho most are doubtless common to both states. Frost's collection, presented to the Brattleboro library, is not accessible at present. Latterly a large amount of work on Vermont fungi has been done by Professor Burt, and Professor L. R. Jones has recently organized a systematic survey to investigate the cryptogamic flora of the state. The following papers pertain to the local flora:

Burt. A List of the Vermont Helvelleae with descriptive Notes. Rhodora, 1: 59–67. *Pl. 4.* 1899.

Frost. Further Enumeration of New England Fungi. Proc. Boston Soc. Nat. Hist. 12: 77–81. 1869.
A continuation of Sprague's list giving 263 species. (See under Massachusetts.)

——— Catalogue of Boleti of New England, with Descriptions of new Species. Bull. Buffalo Soc. Nat. Hist. 2: 100–105. 1874.

Jones & Orton. A partial List of the parasitic Fungi of Vermont. Rep. Vt. Agric. Exper. Sta. 11: 201–219. 1898.
List of 139 Vermont species.

Peck. New Species of Fungi. Bot. Gaz. 5: 33–36. 1880.
Includes ten species from Vermont.

Virginia.

Except some collections made in the vicinity of Washington by the mycologists of the Division of Vegetable Pathology and some papers on various injurious fungi issued from the experiment station little is known of the fungus flora of this state. Incidental species from the state may be found scattered in our literature but no papers bearing on the local flora have appeared.

Washington.

A considerable number of species have been collected by Professor Piper, by W. N. Suksdorf and Miss Adella M. Parker; most of these are represented by duplicates in the Ellis herbarium. A number of new species from this state have been described by Mr. Ellis in his various papers, but only the following papers bearing wholly or in the main on Washington species have appeared:

Ellis & Everhart. New Species of Fungi from Washington Territory by W. N. Suksdorf during the Summer and Fall of 1883. Bull Washburn Coll. Lab. Nat. Hist. 1: 3-6. 1884.

Descriptions of sixteen species.

—————— New West American Fungi. Erythea, 1: 197-206 1893.

Includes several species from Washington.

West Virginia.

Most of our knowledge of West Virginia fungi is due to the work of Mr. L. W. Nuttall, altho some species have been collected through the state Experiment Station. Probably upwards of a thousand species are known from the state, and much of the richest forest region is wholly unexplored. A summary up to 1896 appears as follows:

Millspaugh & Nuttall. Flora of West Virginia. Field Columbian Museum, Bot. Series, 1: 69-276. 1896.

Includes 980 species of fungi. This is a second edition, the first appearing as Bulletin 24, West Virginia Experiment Station.

Wisconsin.

Lapham collected a few species which were sent to Curtis for identification. Later, in 1883, W. F. Bundy published a list containing principally fleshy and woody conspicuous species, but unfortunately there is nothing to stand sponsor for this collection which renders it practically worthless. Later Professor Trelease published a list of parasitic species and collected largely in other groups. These collections are at the Missouri Botanic Gardens. Later collections were made by A. B. Seymour, L. S. Cheney, and particularly Dr. J. J. Davis, of Racine, who has published a supplemental list of parasitic species bringing the number up to five hundred. Special papers bearing on this flora are as follows:

Bundy. A partial List of the Fungi of Wisconsin with Descriptions of new Species. Geology of Wisconsin, 1: 396–401. 1883.

Includes a list of over three hundred species, with two new species.

Davis. Two Wisconsin Fungi. Bot. Gaz. 19: 414, 415. 1894.

———— A graminicolous Doassansia. Bot. Gaz. 25: 353, 354 1898.

———— A supplementary List of parasitic Fungi of Wisconsin. Trans. Wis. Acad. Sci. 9: 153–188. 1892.

Includes 233 species additional to Trelease's list with notes.

Trelease. Preliminary List of the parasitic Fungi of Wisconsin. Trans. Wis. Acad. Sci. 6: 106–144. 1886. (Separate, pp. 40.)

Includes 271 species and descriptions of sixteen new species.

———— The Morels and Puffballs of Madison. Trans. Wis. Acad. Sci. 7: 105–120. *Pl. 7–9.* 1889.

Wyoming.

Comparatively little has been done in the direction of local study of fungi in this state. Professor Aven Nelson has collected a few species in connection with his work on the higher flora of the state, and Professor T. A. Williams and David Griffiths have collected largely the parasitic fungi in connection with other work. Most of these collections are as yet unstudied. The only paper bearing on the local flora pertains equally well to the flora of the states of Montana and South Dakota.

Griffiths. Some Northwestern Erysiphaceae. Bull. Torrey Bot. Club, 26: 138–144. 1899.

* * *

In order to supplement the foregoing, since our southern flora partakes in part of adjoining lands and our northern flora is not limited by the artificial boundaries of lakes and rivers and parallels of latitude, we include a brief statement of our knowledge of the fungi of other portions of North America beyond the limits of the United States.

Canada.

A few of the early northwest passage explorers collected fungi from the Arctic regions and carried them to England. Some of them found their way to Hooker's collection (now at Kew) and were described by Berkeley. Later collections have been made by John Macoun, the Dominion botanist, by John Dearness, of Ontario, and Rev. A. E. Waghorne, of Newfoundland. Most of these have been sent to Mr. Ellis for determination so that duplicates are in his herbarium. Few papers have yet appeared bearing on the fungi of the Dominion, but many species exist in collections and some descriptions are scattered widely through our literature.

The following bear directly on the Canadian flora:

Berkeley. Descriptions of exotic Fungi in the Collection of Sir W. J. Hooker, etc. Ann. Nat. Hist. 3: 375-401. 1839; 7: 451-454. 1841.

———— Enumeration of the Fungi collected during the Arctic Expedition, 1875-76. Jour. Linn. Soc. 17: 13-17. 1880.

Ellis & Everhart. Canadian Fungi. Jour. Mycol. 1: 85-87. 1885.

List of thirty-four species with descriptions of five new species.

Ellis & Dearness. New Species of Canadian Fungi. Can. Record Sci. —: 267-272. 1893.

Descriptions of twenty-one species.

Macoun. List of Plants collected on the Coasts of Labrador, Hudson's Strait and Bay, by Robert Bell in 1884. Rep. Geol. Nat. Hist. Survey, Canada, 1882-1884. App. DD. 38-47. 1885.

List of ten fungi.

Thümen. Contribution a la flore mycologique de la province de Quebec. Nat. Canadien, 10: 8-10. 1878.

Greenland.

Naturally the investigation of the fungi of this Danish possession should be looked for in Danish publications. A summary of the knowledge of Greenland fungi appears in the following paper:

Rostrup. Fungi Groenlandiae: Oversigt over Grönlands Svampe. Meddelelser om Grönland, 3: 517-590. 1888.

Notes on 290 species including several new species.

Mexico.

Our knowledge of the fungi of Mexico is very limited. Several papers appeared early bearing on the fungi of this region obtained from various collectors. More recently collections have been made by T. S. Brandegee, and E. W. Holway; the latter has made several trips seeking especially Uredinales. As in all the countries of Spanish settlement, the work of collecting and describing the lower plants of Mexico will devolve on the Anglo-Americans, and long before the mycological flora of the United States is thoroly known, that of the tropical regions to the southward will be collected by American botanists and the progress in our knowledge of all North America will become more and more uniform. We can refer to the following papers bearing on the mycological flora of Mexico:

Berkeley. On some new Fungi from Mexico. Jour. Linn. Soc. 9 : 423-425. *Pl. 12.* 1867.

Describes six species.

Ellis & Everhart. New West American Fungi. Erythea, 5 : 5-7. 1897.

Descriptions of eight species, six from lower California.

Fries. Novae symbolae mycologicae in peregrinis terris a botanici Danicis collectae. Nova Acta Sci. Upsal. 1 : 17-136. 1851.

——— Novarum symbolarum mycologicarum mantissa. Nova Acta Sci. Upsal. 1 : 225-231. 1851.

Includes a number of species from Mexico.

Holway. Mexican Fungi. Bot. Gaz. 24 : 23-38. 1897.

Forty-seven new species mostly Uredinales.

Central America.

Most of our knowledge of the fungi of Central America has been derived from the explorations of a single expedition fitted out from the Iowa State University ; the results have not all been published but the following have appeared ; duplicates of a considerable part of this collection are in the Ellis herbarium :

Ellis & Everhart. New Species of tropical Fungi. Bull. Lab. Nat. Hist. State Univ. Iowa, 4 : 67-72. 1896.

Descriptions of eleven Nicaraguan and two Mexican species.

Macbride. Nicaraguan Myxomycetes. Bull. Lab. Nat. Hist. State Univ. Iowa, 3 : 377-383. *Pl. 10.* 1893.

——— An interesting Nicaraguan Puff-Ball. Bull. Lab. Nat. Hist. State Univ. Iowa, 3: 216, 217. 1895.

Macbride & Smith. The Nicaraguan Myxomycetes with Notes on certain Mexican Species. Bull. Lab. Nat. Hist. State Univ. Iowa, 4: 73–75. 1896.

Smith. Some Central American Pyrenomycetes. Bull. Lab. Nat. Hist. State Univ. Iowa, 3: 394–415. 1893.

Notes on sixty-one species of which one third or more are new.

West Indies.

A few species were early collected in Jamaica, and Swartz (1806) refers to some nineteen species known to him. Botanists of various countries owning possessions in the various islands have collected a few species here and there. Charles Wright in his expedition to eastern Cuba made the most extensive collection that has ever been made; others less extensive are noted in the literature below. The most recent collections have been made in Porto Rico by Mr. A. A. Heller sent out under the auspices of the New York Botanical Garden; his collections have not yet been studied. The following literature bears on various portions of the archipelago:

Berkeley. Notices of Fungi in the Herbarium of the British Museum. Ann. Mag. Nat. Hist. 10: 369–385. *Pl. 9–12.* 1843.

Includes species from Jamaica.

——— Enumeration of some Fungi from St. Domingo. Ann. Mag. Nat. Hist. II., 9: 192–203. *Pl. 8.* 1852.

List of sixty-seven species including descriptions of twenty new species.

Berkeley & Curtis. Fungi Cubenses. Jour. Linn. Soc. 10: 280–341, 341, 392. 1869.

List of numerous species from Cuba with descriptions of many new species.

Bresadola, Hennings, & Magnus. Die von Herrn P. Sintenis auf der Insel Porto Rico 1884–1887 gesammelten Pilze. Engler's Jahrb. 17: 489–501. *Pl. 12.* 1893.

Notes on sixty-four species including six new species.

Cockerell. Notes on some Fungi collected in Jamaica. Bull. Torrey Bot. Club, 20: 295–297. 1893.

Notes on twenty-four species.

Cooke. Some exotic Fungi. Grevillea, **17** : 59–60. 1889.
Descriptions of five new species.

Ellis & Kelsey. New West Indian Fungi. Bull. Torrey Bot. Club, **24** : 207–209. 1897.
Descriptions of six species from St. Croix.

Hennings. Fungi jamaicenses. Hedwigia, **37**: 277–282. 1898.
Thirty-three species, twelve new.

Hitchcock. List of Cryptogams collected in the Bahamas, Jamaica and Grand Cayman. Rep. Mo. Bot. Gard. **9**: 111–120. 1898.
Thirty-one species mentioned, nine new.

Massee. Some West Indian Fungi. Jour. Bot. **30**: 161–164, 196–198. *Pl. 321–323, 325.* 1892.
List of ninety-one species including eighteen species new.

Montagne. Troisième centurie de Plantes cellulaires exotiques nouvelles. Fungi Cubensis, I. Ann. Sc. Nat. II. **17**: 119–128. 1842.
Forty new species from Cuba.

Rostrup. In Borgesen og Paulsen : Om Vegetation paa de danskvestindiske Oer. Bot. Tidsskr. **22**: 110–112. 1898.
List of thirty-one species from the Danish West Indies.

Roussel. Énumération des Champignons récoltés par M. T. Husnot aux Antilles françaises en 1868. Bull. Soc. Linn. Norm. Caen. II. **4**: 217–225. 1868–9.

Sagra. Icones plantarum in Flora Cubana. Folio, Paris, 1863.
List of 116 species of fungi; pp. 47, 48 and plates 11–17 illustrate Cuban species.

Swartz. Flora Indiae Occidentalis. **3**: 1920–1939. 1806.
Includes nineteen species of fungi from Jamaica.

It will be apparent to those who consider carefully the extensive areas of our own country where mycological exploration has scarcely been made, that at the threshold of the twentieth century, we have really made only a beginning in the study of the extent and distribution of the fungus flora of North America.

CHAPTER XII

METHODS OF COLLECTION AND PRESERVATION OF FUNGI

HINTS FOR FURTHER STUDY

The collection and preservation of material is an important factor connected with the study of fungi as of other plants. Specimens are of value as a permanent record of the nature and distribution of a species according as they are (1) Collected in abundance and in a mature condition, (2) Carefully preserved, and (3) Accessible for reference in a public collection, or one that can be readily consulted. The mere statement that this or that fungus grows in a certain part of the country is of value only as one has confidence in the ability of the person making the statement, to properly identify the species in question ; a well preserved and accessible specimen stands as a permanent voucher for the statement and in addition shows the character of the species. It is desirable, therefore, that material (1) Be collected with care and such field notes be taken as will supplement its characters, (2) Be preserved in such a way that the essential characters will suffer the least possible injury, and (3) Be deposited in a public collection where it will be properly cared for and be accessible to workers who are studying the character and distribution of the species to which it belongs. The curse of much of the early systematic study of American fungi is found in the miserable specimens, defective both in the quantity of material preserved and in its proper maturity, that in too many cases have served as the types of described species. The detriment to much of the recent systematic study is the inroads on the time of monographers who are besieged to name this or that lot of species, for anxious collectors.

The forms of fungi are so diverse that no general directions for collection and preservation can be given that will apply in all cases. Some can be best preserved dry, some pressed as herbarium specimens, some in fluid, and some as microscopic preparations, the purpose being to retain in the preserved specimen just as many of the essential characters of the fresh material as possible.

Parasitic species on leaves or stems are simply preserved by ordinary pressure. In collecting these, as all fungi growing on plants living or dead, the greatest care should be taken to determine the host on which the fungus grows. If not already a familiar plant, the buds, flowers, fruit or such other available evidence should be collected together with the leaves, as will enable some one to recognize the host specifically. In the case of fungi growing on stumps, fallen logs or on dead branches this is not always so easy, yet bark, or other accessary data can often be obtained by the field collector that will render reasonably certain the genus, if not the species, of the tree on which the fungus grew.* This involves, on the part of the collector, a knowledge of the higher plants, and for one to be really successful, the wider this acquaintance the better. A really good specimen of a leaf-inhabiting fungus ought to consist of at least a dozen well-affected leaves. A specimen of a fleshy, woody or leathery fungus ought to contain from a half dozen to a dozen individuals, if possible, in different states of development; in many cases a larger quantity of material is desirable, and any sample can far better possess too many individuals than too few.

Fleshy ascomycetous fungi (Helvellales, Pezizales, etc.) can often be best preserved in alcohol, tho many of them with proper care can be dried so as to be quite satisfactory.

Agarics and *Boleti* can be best preserved by drying in a current of hot air suspended in a wire tray over a gas burner or even over an oil lamp. In collecting fleshy forms, a basket can well be used for carrying specimens, but each form should be carefully wrapped in tissue paper so as to prevent breaking, marring, or soiling from contact with other specimens.† Careful field notes should

---

* In reporting hosts the Latin name rather than the English should be given. To say, for example, that a fungus grows on a poplar, is indefinite, since a poplar in the East is of the genus *Populus* while in the central West it is a *Liriodendron*, a wholly different tree.

† A very valuable paper has been prepared by Professor Burt on methods of collecting fleshy fungi. *Cf.* Burt, On Collecting and Preparing fleshy Fungi for the Herbarium. Bot. Gaz. **25**: 172–186. *Pl. 14.* 1898.

be taken, the nature of which will vary somewhat in the various groups represented.

In collecting any fleshy fungi, care should be taken to obtain all the fleshy structure, because some very important characters are derived from the basal parts. They should never be gathered for scientific purposes by breaking them off above the ground. The entire basal portion should be removed with a knife or small trowel. Of course the date of collecting and the locality will be added to the specimen by any intelligent collector, but it is always desirable to add the local environment, of the specimen by stating in what soil it grows—sand, clay or leaf-mold—and whether the plant grows in open pastures, marsh, grassy woods, or deep forest; sometimes the character of the timber, especially pine land is to be noted, also whether the fungus grows singly or in clusters. But above all these matters of environment, certain data concerning the physical properties of the fresh plant are absolutely essential to a correct understanding of the species. Dried specimens of fleshy fungi without notes are often worse than useless for they suggest many times highly interesting and often undescribed species without sufficient data to enable one to characterize them properly, and, ordinarily, the species of fleshy fungi had better be left undescribed than be named exclusively from the dried plant. The summary of characters to be noted in the Agaricaceae may be tabulated as follows:

1. TASTE.—Bitter, acrid, peppery, mealy, nutty? (One need feel no fear in tasting any of the fleshy fungi for they are cleanly, and the only inconvenience ever experienced is the peppery taste of certain species of *Lactarius* and *Russula* which is temporarily about as unpleasant as tasting a particle of red pepper, but otherwise harmless.)

2. SURFACE OF PILEUS. — Dry, hygrophanous or viscid? Smooth, granular, scaly, shining, striate, unbonate, umbilicate? Color and size?

3. LAMELLAE.—Color when young, and when mature? Close or distant? Narrow or wide? Entire, heterophyllous, or anastomosing? Decurrent on the stem, adnate, sinuate, or free?

4. SPORES.—(Best collected by removing the pileus and placing it lamellae downward on paper or glass under a tumbler or bell jar. If a microscope is at hand to examine the spores they can be best collected on a slide.) Color, shape and size?

5. STEM.—Fleshy throughout or with a cartilaginous rind? Hollow, solid, or stuffed? Size, including length and thickness? Shape; cylindric, tapering, radicate, or bulbous?

6. VOLVA and VEIL if present; character and position?

To these notes a simple sketch of the fully expanded plant, together with earlier stages, preferably in colors, will be a very valuable addition.

The specimens should be dried as quickly as possible after being collected, as they are the favorite food of certain insect larvae, and if left over night will often be found to have changed into disgusting heaps of corruption by morning.

In the Boletaceae the color of the spores should be determined in the manner indicated for the agarics, and the taste of the fresh specimen is also essential. In addition, the colors of the pileus, flesh, and pores should be noted, and if there is any difference in color between the young pores and those of the mature plant, this fact should be noted also. In certain species the flesh or pores or both, will change color rapidly or slowly when wounded; in some instances the change is to a bright blue; this changing condition should be noted in any given species. Any peculiarity of shape of stem or markings on the stem like veining, reticulation, or glandular dots should be carefully noted. If a veil is present, its character will be important, as well as the relation of the pores themselves to the stem, whether adnate, free, or merely depressed around it. Finally the character of the pileus should be noted, whether viscid or dry. Specimens need to be dried rapidly, and after the drying has once commenced, it should be carried to the end without stopping.

In the Clavariaceae the color of the spores, taste, odor, and color of the fresh plant should be carefully noted, as well as the characters of the tips of the branches.

All members of the order Phallales should be preserved in alcohol (60-70%); it is especially desirable that the earliest stages up to the so-called eggs—should be preserved when possible, in order to make possible a more complete study of their development.

The Lycoperdaceae and other puff-balls should be preserved dry; if possible the specimens should retain their outer peridia, and young forms should not be neglected; they should invariably be preserved in boxes without pressure.

The specimens which are flat and not likely to be injured by rubbing, may either be pasted directly to herbarium sheets* or preserved in pockets or envelops of which several sorts are to be had of dealers in botanical supplies, or one can easily fold them. The best for most forms of plants are those which fold under at least three quarters of an inch at each end since they securely hold the specimen in place when attached to the herbarium sheet by a drop of glue. Bulky specimens like most agarics,† woody and fleshy fungi like *Polyporus* and *Boletus*, puff-balls, morels, many forms of *Hypoxylon*, *Xylaria*, Thelephoraceae, and the like, can be best preserved in boxes.‡

Myxomycetes and some of the collapsible moulds (*Moniliales*, etc.) require to be glued direct to the box cover so that they will not be injured by handling.§

Too much care can hardly be given to all these details, for a

---

* It is usually better, however, to paste first to a small sheet which may later be attached to the herbarium sheet and thus prevent its wrinkling.

† Some prefer to press these lightly. This can easily be done in the smaller and thinner species. After they are once fully dry wrap for a few moments in a moist cloth when they will become pliable. Only a light pressure should be applied.

‡ The writer has found convenient a series of multiple sizes of pasteboard boxes, and the same forms have been adopted by the New York Botanical Garden. The most useful sizes are as follows:

	Length.	Width.	Depth.
1	2¼	2	1¼
2	4	2¼	1¼
3	4	2¼	2½
4	5½	4	1¼
5	5½	4	2½
6	8	5½	2½

By using these multiple sizes a closer packing is possible. The multiples are themselves fractions of a standard herbarium sheet so as to render possible an arrangement in sequence in a herbarium if desired.

§ For material of this sort the writer uses a special form of box ¾ inch deep with cover 4 × 1¾ inches. By gluing the specimen to the cover it cannot become separated from the label which is most conveniently written on the cover itself, and the specimen can be easily examined *in situ* under a dissecting glass or even under the lower powers of a compound microscope.

well-preserved collection, if properly prepared in the first place, and properly cared for in the second, will outlast several generations of botanical workers.

**

After a student has become familiar with a series of typical fungi,* the preceding pages should enable him to locate the specimens of the more conspicuous saprophytes and parasites he may collect, as far as the genus, and in the case of ordinary edible forms as far as the species. But the ambitious student will not stop here; for this reason the leading systematic papers, many of which can be consulted only in the larger libraries or botanical laboratories, have been freely referred to in the text. It will probably be some years before we will have descriptive manuals for our fungi as for our higher plants, and their value when prepared will depend in no small degree on the careful notes taken by the individual workers all over the country, provided the results of their work are properly vouched for by carefully preserved material deposited in one or more of the great botanical centers where future systematic studies will be largely carried on. Until we can have manuals of our own we must depend for the determination of species † on Saccardo's *Sylloge Fungorum* and the references in its successive supplements. For those who read German, the portions of Rabenhorst's Kryptogamen-Flora von Deutschland, Oesterreich, und der Schweiz, relating to Die Pilze, will be valuable particularly for the Phycomycetes and Ascomycetes; Massee's British Fungus Flora (4 vols.) and Stevenson's British Fungi (2 vols.) will be useful for many of the Basidiomycetes especially. In all these European manuals, however, it must be remembered that only a portion of our flora is in common with that of northern Europe and many of our species have no place at all in the European manuals.

But the name of the fungus and its position in the system are only means to a further end. While a familiarity with plants is

---

* This series ought at least to include the common black and green moulds, a powdery mildew like that on the lilac or willow, a rust, a mushroom, a puff-ball, a cup-fungus, and a *Xylaria* or other of the Sphaeriales.

† The student of parasitic fungi will find the following work valuable: **Farlow & Seymour**, A provisional Host Index of the Fungi of the United States. Pp. 218. Cambridge, 1888–1891.

one of the most important qualifications of a botanist, and one too often neglected in these later days, it is a foundation merely to something better. After something of a familiarity with fungi is gained, DeBary's Morphology of the Fungi, which to a novice would be dry and unintelligible, will become interesting and valuable, and should serve as a work of constant reference. According to the taste of the individual, studies may be prosecuted in the life history of certain species, in the development and relations of the various stages of polymorphic species,* in pure cultures of isolated imperfect forms or of mycelia now known only in a sterile condition, in the physiology of growth or nutrition, in the ecologic relations of fungi, in the study of certain species injurious to vegetation, or in further systematic revision, a field which sadly needs the most efficient workers, who can combine with the study of herbarium specimens and literature, extensive field study, and long continued cultivation in the fungus garden and the Petri dish. The best taxonomic work of the future must involve both morphological and physiological study together with extended cultures; a few fields have been comparatively well worked over, but the many await the future monographer.

---

* Bearing on the subject of culture media and culture methods a student can profitably read, **Smith.** Hints on the Study of Fungi. Asa Gray Bull. 4: 25-28; 37-43. 1896. This was prepared by one of our most careful experimental investigators and contains valuable suggestions for one form of mycological investigation. As a suggestion to a beginner, however, it is an educational curiosity, and would in no way serve as a help to the honest seeker after a little truth.

ADDITIONS AND CORRECTIONS

Page 15, line 19; for *f. 3*, read *f. 6*.
Page 23, line 26; for *Pl. 2*, read *Pl. 3*.
　　　　line 27; for *Pl. 2*, read *Pl. 3*.
Page 27, line 7 from bottom : insert 4 : before 1881.
Page 38; *dele* line 7 from bottom.
Page 63, line 10 from bottom; for are, read is.
Page 78, line 16; for from, read form.
Page 85, line 18; for 11 and 12 : 1895 read, 5 : 1–220. *Pl. 1–13*. 1883; 11 : 1–98. *Pl. 1–5*. 1895; 12 : 99–236. *Pl. 6–12*. 1895.
Page 92, line 4; *dele* * after PHRAGMIDIUM.
Page 93. To the literature should be added :
Carleton. Studies in the Biology of the Uredineae. I. Notes on Germination. Bot. Gaz. **18** : 447–457. *Pl. 37–39*. 1893.
Page 106, line 5; for or, read on.
Page 109 ; insert the words reddish-brown before pileus (line 13).
Page 122, line 5 from the bottom should commence with the word it.
Page 124, first line of first footnote ; for specimens, read species.
Page 131. To the literature add :
Cooke. Illustrations of British Fungi, 8 vols. *Pl. 1–1198*. London, 1881–1891.
Page 159, line 12 ; for 1897, read 1895.
Page 197. To the literature bearing on Greenland add :
Rostrup. Tillaeg til Grönlands Svampe (1888). Meddelelser om Grönland, **3** : 591–643. 1891.
Increases the Greenland list to 532 species.

INDEXES.

I. Index to Latin Names

Note.—Generic names are in Roman, families in SMALL CAPITALS, orders in **bold face**, classes and higher groups in CAPITALS, synonyms or obsolete names in *italics*. Families and orders can also be recognized by their terminations.

Abrothallus, 61
Absidia, 26
Acetabula, 56
Acrasiales, 149
ACROSPERMACEAE, 52
Actinonema, 69, 70
Aecidium, 86, 92, 157
AGARICACEAE, 98, 99, 105, 109, 203
Agaricales, 80, 95, 96, 97, 129
Agariceae, 119
Agaricus, 19, 98, 109, 114, 122, 130, 144, 155, 157
Agyrium, 60
ALBUGINACEAE, 32
Albugo, 32
Aleuria, 56
Aleurodiscus, 100, 101
Alternaria, 77
Amanita, 19, 110, 112, 119, 120, 155, 157
Amanitopsis, 112, 120
Amaurochaete, 150
Ampelomyces, 69
AMPHISPHAERIACEAE, 48
Anellaria, 114, 122
Angioridium, 151
Annularia, 113, 121
Annulatae, 121
Anthracoidea, 83
Anthracophyllum, 114, 118
Anthurus, 132, 133
Arachnion, 143
Arachnopeziza, 58
ARCHEGONIATA, 8, 9
Archimycetes, 22

Arcyria, 152, 157
Aregma, 91
Armillaria, 112, 121
Arrhytidia, 96
Artotrogus, 29
ASCOBOLACEAE, 55, 56
Ascobolus, 57, 157
Ascochyta, 69, 70
ASCOCORTICIACEAE, 38
Ascocorticium, 38
ASCOMYCETES, 18, 34
Ascophanus, 57
ASPERGILLACEAE, 40
Aspergillales, 36, 39
Asterostroma, 101
Astraeus, 138, 140
Auricularia, 94, 156
AURICULARIACEAE, 94
Auriculariales, 80, 81, 94

Bactrospora, 61
Badhamia, 150
Baggea, 61
BALSAMIACEAE, 51, 52
BASIDIOMYCETES, 18, 80
Basidiophora, 32
Batarrea, 138, 139, 140, 157
Belonidium, 60
Beloniella, 59
Belonioscypha, 58
Belonium, 58
Belonopsis, 60
Biatorella, 61
Bolbitius, 113, 116
BOLETACEAE, 98, 99, 106, 204

Boletinus, 106, 107, 130
Boletus, 98, 107, 108, 109, 130, 155, 157, 205
Bostrichonema, 76
Botrytis, 75, 76, 157
Boudiera, 57
Bovista, 136, 139, 141
Bovistella, 139, 141
Bremia, 32, 33
Bulgaria, 62
Bulgariella, 62

Caeoma, 92
Caldesia, 60
Caeomurus, 91
Calloria, 60
Calocera, 96
Calonema, 152
Calostoma, 138, 139, 141
Calvatia, 136, 139, 141
Calyptospora, 90
Camarosporium, 70
Campsotrichum, 76
Cantharelleae, 115
Cantharellus, 101, 113, 115, 130, 155
Capnodium, 77
Caryospora, 47
Catastoma, 137, 139, 141
Cauloglossum, 139, 140
CELIDIACEAE, 55, 60
Celidium, 60
CENANGIACEAE, 55, 61
Cenangiella, 61
Cenangium, 61
Cenococcum, 143
Ceracea, 96
Ceratiomyxa, 149
Ceratophorum, 77
CERATOSTOMACEAE, 48
Cercospora, 74, 77, 78, 79
Cercosporella, 76
Chaetocladium, 25, 27
CHAETOMIACEAE, 48
Chaetomella, 70
Chitonia, 114, 120
Chlorosplenium, 57, 58
Choanophora, 27
CHOANOPHORACEAE, 27
Chondrioderma, 150
Chrysomyxa, 89, 90

Chytridiales, 22, 23, 152
Ciboria, 58
Cidaris, 65
Cienkowskia, 151
Cintractia, 83
Cladosporium, 75, 76
Clarkeinde, 120
Clasterosporium, 77
Clastoderma, 150
CLATHRACEAE, 132
Clathroptychium, 151
Clathrus, 132, 133, 155, 157
Claudopus, 112, 125
Clavaria, 96, 98, 102, 130, 155, 157
CLAVARIACEAE, 63, 98, 101, 204
Claviceps, 43, 158
Clitocybe, 112, 124
Clitopilus, 113, 126, 127
CLYPEOSPHAERIACEAE, 48
Coleosporium, 89, 90
Colletotrichum, 73
Collybia, 17, 112, 124, 128
Collyria, 96
Comatricha, 151
Conida, 60
Coniophora, 100, 101
Coniothyrium, 70
Coniosporium, 76
Coprinarius, 124
Coprineae, 115
Coprinus, 114, 115, 116, 130
CORDIERITHACEAE, 55
Cordieritis, 55
Cordyceps, 39, 44, 78, 155, 158
Cornuella, 84
Corticium, 100, 130
Cortinarius, 110, 113, 123
Coryne, 59
CORYNELIACEAE, 48
Coryneum, 73
Craterellus, 100, 101, 115, 130
Craterium, 150
Crepidotus, 112, 124, 125
Cribraria, 151
Cronartium, 89, 90
Crucibulum, 142
Crumenula, 61
Cryptosporium, 73
Cryptostictis, 70
Cubonia, 56

CUCURBITARIACRAE, 48
Cudonia, 64
Cudoniella, 64
Cyathicula, 58
Cyathus, 142, 157
Cyclomyces, 105
Cylindrosporium, 73
Cyphella, 100
Cystopus, 32
Cytidium, 150
CYTTARIACEAE, 54
Dacryomyces, 96
Dacryomycetales, 80, 96
Dacryopsis, 96
Daedalea, 105, 130, 155, 157
Daldinia, 49
Darluca, 70
Dasyscypha, 58, 59
Deconica, 114, 125
DEMATIACEAE, 75, 76
Dermatea, 62
Derminus, 124
Desmazierella, 58
Diachaea, 150
DIATRYPACEAE, 49
Dicaeoma, 91
DICHAENACEAE, 52
Dicranophora, 26
Dictydium, 151
Dictyophora, 133, 134
Diderma, 150, 157
Didymaria, 76
Didymium, 150
Dinemosporium, 72
Diplodia, 69
Discina, 56
Discosia, 71
Ditiola, 96
Doassansia, 84, 85
Doassansiopsis, 84, 85
DOTHIDEACEAE, 45
Dothideales, 36, 45
Durella, 60

Eccilia, 113, 126, 127
Elaphomyces, 39, 44
ELAPHOMYCETACEAE, 40
Ellisiella, 76
Empusa, 28
ENDOMYCETACEAE, 37
Endophyllum, 89

Enerthenema, 150
Entoloma, 113, 127
Entomophthora, 28
Entomophthorales, 22, 23, 28
Entomosporium, 20, 71, 72
Entyloma, 84
Epichloe, 44
Erinella, 58
Eriopeziza, 58
ERYSIBACEAE, 42, 69, 76.
Erysibe, 42, 156
Erysiphe, 42
Excipula, 72
EXCIPULACEAE, 69, 72
Exidia, 95
EXOASCACEAE, 38
Exoascales, 35, 37
Exoascus, 38, 97
EXOBASIDIACEAE, 97
Exobasidiales, 80, 97
Exobasidium, 18, 97
Evelatae, 124, 127

Fabraea, 59
Favolus, 105
Flammula, 113, 124
Fomes, 105
Frankia, 149, 153
Fuligo, 150, 157
Fumago, 77
Fusarium, 75, 78
Fusicladium, 75, 76

Galactinia, 56
Galera, 114, 124, 126
Gasteromycetes, 80
Gautieria, 135
Geaster, 138, 140, 157
GEOGLOSSACEAE, 63
Geoglossum, 64, 157
Geopora, 52
Geopyxis, 56
Glenospora, 77
Gloeopeziza, 57
Gloeoporus, 104, 105
Gloeosporium, 73
Glomerularia, 75
GNOMONIACEAE, 48
Godronia, 61
Gomphidius, 114, 116
Gorgoniceps, 59

Grandinia, 103, 130
Guepinia, 96
GYMNOASCACEAE, 39
Gymnoconia, 91, 92
Gymnosporangium, 18, 87, 91, 156
Gyrocephalus, 95
Gyromitra, 65, 66
Gyrophragmium, 130, 138, 140

Hadrotrichum, 76
Haematomyces, 62
Haematomyxa, 62
Hainesia, 73
Hebeloma, 114, 124
Heliomyces, 113, 118
Helminthosporium, 74, 77
HELOTIACEAE, 55, 57
Helotium, 57, 58, 59, 157
Helvella, 65, 66, 155, 157
HELVELLACEAE, 63, 65
Helvellales, 36, 63, 202
Hemiarcyria, 152
Hendersonia, 70
Herpocladiella, 27
Heterosporium, 77
Heterotrichia, 152
Hexagonia, 105
Hiatula, 112, 128
Holwaya, 62
Hormomyces, 96
Humaria, 56
HYDNACEAE, 98, 99, 102
Hydnum, 19, 95, 98, 102, 103, 104, 130, 155
Hygrophoreae, 116
Hygrophorus, 112, 116
Hymenochaete, 101
Hymenogaster, 135
HYMENOGASTRACEAE, 135
Hymenogastrales, 81, 135
Hymenomycetes, 80, 97
Hymenoscypha, 58, 59
Hypholoma, 114, 123
HYPOCHNACEAE, 98, 99
Hypochnus, 99
HYPOCREACEAE, 45
Hypocreales, 36, 43
Hypocreella, 44
HYPODERMATACEAE, 52, 53
Hypomyces, 44
Hyporhodius, 124

Hypoxylon, 47, 49, **205**
Hysterangium, 135
HYSTERIACEAE, 52, 53, 72
Hysteriales, 36, 152
Hysteropatella, 61

Illosporium, 78
Inocybe, 114, 124
Irpex, 102, 103, 130
Isaria, 78, 157
Ithyphallus, 133

Johansonia, 61

Karschia, 61
Kneiffia, 103

LABOULBENIACEAE, 51
Laboulbeniales, 35, 50
Labrella, 71
Lachnea, 56
Lachnella, 58, 59
Lachnellula, 58
Lachnocladium, 102
Lachnobolus, 152
Lachnum, 58, 59
Lactariae, 116
Lactarius, 112, 116, 117, 118, 155, 203
Lahmia, 61
Lamproderma, 150
Lasiobolus, 57
Laterna, 132
Lecideopsis, 60
Leciographa, 61
Lentinus, 113, 119, 130
Lentodium, 119
Lenzites, 105, 130, 155
Leotia, 64, 157
Leocarpus, 150
Lepidoderma, 150
Lepiota, 15, 112, 121
Leptoglossum, 64
Leptonia, 113, 127
Leptostroma, 71
LEPTOSTROMATACEAE, 69, 71
Leptothyrium, 72
Licea, 151, 157
Limacium, 116
LOPHIOSTOMACEAE, 48
Lycogala, 151, 157
LYCOPERDACEAE, 136, 138, 204

Lycoperdales, 81, 136
Lycoperdon, 130, 139, 141, 155, 156, 157
Lysurus, 133

Macropodia, 56
Macrosporium, 74, 77
Magnusiella, 38
Mamiana, 47
Marasmieae, 118
Marasmius, 113, 118
Marsonia, 73
MASSRAIACEAE, 48
MELAGRAMMATACEAE, 49
Melampsora, 89, 90
MELAMPSORACEAE, 89
MELANCONIACEAE, 72
Melanconiales, 68, 72
MELANCONIDACEAE, 49
Melanogaster, 135
Melasmia, 72
Melaspilea, 61
Meliola, 18, 39
Merulius, 104, 130
Michenera, 100, 101
Microglossum, 64
Microsphaera, 42
Microstroma, 97
MICROTHYRIACEAE, 42, 43
Mitremyces, 139
Mitrula, 64
Mollisia, 60
MOLLISIACEAE, 55, 59
Mollisiella, 60
Monilia, 20, 75, 157
MONILIACEAE, 75
Moniliales, 68, 74, 205
Montagnites, 114, 116, 130
Morchella, 65, 66, 155, 157
Mortierella, 27
MORTIERELLACEAE, 26, 27
Mucedineae, 75
Mucor, 20, 24, 26, 27, 156, 157
MUCORACEAE, 26, 28
Mucorales, 22, 23, 24, 26
Mucronella, 103
Mucronoporus, 106
Mutinus, 133
Mycena, 112, 124, 128
Mycenastrum, 139, 141
MYCETOZOA, 146, 150

MYCOSPHAERELLACEAE, 48
Mylittopsis, 94
Myriadoporus, 106
Myriostoma, 137, 138, 140
MYXOBACTERIACEAE, 152
Myxogastrales, 149
MYXOMYCETES, 19, 24, 146, 205
Myxosporium, 73

Naucoria, 114, 124
Nectria, 45, 78
Nectrioideae, 71
Nesolechia, 60
Nidularia, 142, 158
NIDULARIACEAE, 141
Nidulariales, 81, 141
Niptera, 60
Nolanea, 113, 127
Nummularia, 49
Nyctalis, 113, 116

Octaviana, 135
Odontium, 103
Oidium, 74, 75, 76
Oligonema, 152
Ombrophila, 59
Omphalia, 112, 124, 127
ONYGENACEAE, 40
Oomycetes, 23
Oospora, 75
Ophiotheca, 151
Orbilia, 60
OSTROPACEAE, 52
Otidea, 56
Ovularia, 75

Panaeolus, 114, 124, 125
Panus, 113, 119
Paryphedria, 62
Patellaria, 61
PATELLARIACEAE, 55, 60
Patellea, 60
Patinella, 60
Paxilleae, 115
Paxillus, 106, 113, 115, 130
Penicillium, 16, 39, 156
Peniophora, 101
Perichaena, 151
Peridermium, 18, 90
PERISPORIACEAE, 42, 43
Perisporiales, 36, 40

Peronospora, 32, 33
PERONOSPORACEAE, 32
Peronosporales, 22, 23, 30, 31
Pestalozzia, 73
Pestalozziella, 73
Peziza, 56, 155, 157
PEZIZACEAE, 55
Pezizales, 36, 54, 67, 155, 202
PHACIDIACEAE, 53
Phacidiales, 36, 53
Phacidium, 53
Phacopsis, 60
PHALLACEAE, 132, 133
Phallales, 81, 132, 204
Phallogaster, 132, 133
Phallus, 133, 155, 157
Phellorina, 143
Phlebia, 102, 103
Phleospora, 70
Phlyctospora, 136
Pholiota, 113, 121
Phoma, 69
Phragmidium, 18, 92
Phragmopyxis, 92
Phycomyces, 26
PHYCOMYCETES, 18, 22
Phyllachora, 45
Phyllactinia, 42
Phyllosticta, 69, 70
Physalacria, 102
Physarella, 150
Physarum, 150
Phytophthora, 32
Piggotia, 72
Pilacre, 94
PILACRACEAE, 94
Pilaira, 26
Pilobolus, 25, 26, 27, 157
Pilocratera, 57, 58
Pilosace, 114, 126
PIPTOCEPHALIDACEAE, 27
Piptocephalis, 25, 27
Piricularia, 76
Pirottaea, 59
Pistillaria, 102
Pitya, 58
Plasmodiophorales, 149
Plasmodiophora, 149
Plasmopara, 32
PLEOSPORACEAE, 48
Pleurotus, 112, 113, 124, 125

Plicaria, 56
Plicariella, 56
Pluteolus, 114, 124, 126
Pluteus, 113, 126
Pocillum, 59
Podaxon, 130, 138, 139, 140
Podosphaera, 42
Poikilosporium, 83
Polyplocium, 138, 139
POLYPORACEAE, 98, 99, 104
Polyporus, 98, 105, 106, 130, 155
Polysaccum, 138, 139, 143
Polystictus, 105
Polythrincium, 76
Poria, 105, 106
Poronia, 49
Porothelium, 105
Pragmospora, 61
Protodermium, 149
Protomyxa, 152, 153
Psalliota, 123, 124
Psathyra, 114, 126
Psathyrella, 114, 124, 125
Pseudhydnotrya, 52
Pseudopeziza, 59
Pseudoplectania, 56
Pseudotryblidium, 62
Psilocybe, 114, 126
Psilopezia, 63
Pterula, 102
Puccinia, 87, 88, 90, 91, 92, 157
PUCCINIACEAE, 89, 90
Pucciniastrum, 90
Pulparia, 61
Pycnodon, 103
Pyrenopeziza, 59
Pyronema, 55
PYRONEMACEAE, 55

Queletia, 138, 140

Radulum, 102, 103, 130
Ramularia, 74, 75, 76, 78, 79
Ravenelia, 18, 91, 92
Ravenelula, 60
Reticularia, 151
Rhizina, 63
RHIZINACEAE, 63, 65
Rhizopogon, 135
Rhytidopeziza, 62
Rhytisma, 53

Rhyparobius, 57
Roestelia, 18, 92
Russula, 112, 116, 117, 203
Russulina, 116
Rutstroemia, 58

Saccardia, 42
SACCHAROMYCETACEAE, 37
Saccharomycetales, 35, 36
Sacidium, 71
Saprolegniales, 22, 23, 29
Sarcobolus, 57
Sarcomyces, 62
Sarcoscypha, 9, 57
Sarcosphaera, 56
Sarcosoma, 62
SCHIZOMYCETES, 19, 152, 154
Schizophylleae, 118
Schizophyllum, 113, 118
SCHIZOPHYTA, 19
Schizonella, 82, 83
Schweinitzia, 61
Scleroderma, 143, 157
SCLERODERMATACEAE, 143
Sclerodermatales, 81, 143
Sclerogaster, 135
Sclerospora, 32
Sclerotinia, 58
Scolecotrichum, 76
Scutula, 61
Scutularia, 61
Scyphium, 150
Sebacina, 95
Secotium, 130, 138, 140
Septogloeum, 73
Septoria, 69, 70
Simblum, 132, 133
Siphoptychium, 151
Sistotrema, 103, 157
Solenia, 100
SORDARIACEAE, 48
Sorokina, 62
Sorosporium, 83
Sparassis, 102
Spathularia, 64
SPERMAPHYTA, 8, 9
SPHAERIACEAE, 48
Sphaeriales, 17, 36, 46, 67, 69, 70
Sphaerobolus, 142, 157
SPHAEROPSIDACEAE, 69, 71
Sphaeropsidales, 68, 69

Sphaeropsis, 69, 70
Sphaerospora, 56
Sphaerotheca, 42
Sporodinia, 25, 26
Spumaria, 150, 157
Stagnospora, 70
Stammaria, 59
Starbaeckia, 60
Stemonitis, 148, 151, 157
Stella, 143
Stereum, 98, 100, 101
STICTIDACEAE, 53
Stigmina, 77
STILBACEAE, 75, 78
Streptotheca, 57
Streptothrix, 77
Strobilomyces, 106, 107, 130
Stropharia, 114, 122, 124
Syncephalis, 27
Syncephalastrum, 27
Synchytrium, 11, 17, 23, 24

Tapesia, 59
Taphria, 38
Taphrina, 38
Tazetta, 56
TERFEZIACEAE, 40
THALLOPHYTA, 8, 9, 19
Thamnidium, 26, 27
Thecaphora, 83
Thelebolus, 142, 157
Thelephora, 99, 100, 101, 130, 157
THELEPHORACEAE, 98, 99, 100, 205
Tilletia, 83, 84
TILLETIACEAE, 82, 83
Tilmadoche, 150
Tolyposporella, 83
Tolyposporium, 83
Tomentella, 99, 100, 130
Torula, 76
Trametes, 105, 106
Tremella, 95, 155, 156, 157
TREMELLACEAE, 95
Tremellales, 62, 80, 81, 95
Tremellodon, 95
Trichia, 152, 157
Trichobelonium, 59
TRICHOCOMACEAE, 40
Tricholoma, 112, 124, 128, 155
Triphragmium, 91, 92
Trochila, 53

Trogia, 113, 115
TRYBLIDIACEAE, 53
Tryblidiella, 62
Tubaria, 113, 124
Tuber, 52, 155, 157
TUBERACEAE, 51, 52
Tuberales, 35, 51
Tubercinia, 84
Tubercularia, 78, 157
TUBERCULARIACEAE, 75, 78
Tuberculina, 78
Tubulina, 151, 157
Tylostoma, 138, 140, 157
Tympanis, 62
Typhula, 102

Ulocolla, 95
Uncinula, 42
Underwoodia, 65
Uredinales, 80, 81, 85
Uredinopsis, 89, 90
Uredo, 86, 92, 157
Urocystis, 83, 84
Uromyces, 18, 87, 88, 90, 91, 92

Uropyxis, 91, 92
USTILAGINACEAE, 82
Ustilaginales, 24, 80, 81
Ustilago, 83, 84
Ustulina, 49

VALSACEAE, 49
Velatae, 123
Velutaria, 61
Vermicularia, 69, 157
Verpa, 65
Vibrissea, 64
Volvaria, 113, 120
Volvatae, 119

Xerotus, 113, 118
Xylaria, 49, 205
XYLARIACEAE, 49

Zukalina, 57
Zygomycetes, 22
Zythia, 71
ZYTHIACEAE, 69, 71

II. INDEX TO HOST PLANTS OF FUNGI

NOTE.—Hosts are here indexed as they appear in the text; cross references are given in case a host is referred to by both its English and Latin name at different pages.

Abies, 90
agarics, 95
Agrimonia, 90
Alisma, 85
Alnus, 62, 149
Amarantus, 32
Amorpha, 92
Andromeda, 53
Anemone, 11, 23
apple, 1, 8, 69, 73, 75, 85, 87
Aralia, 92
Arundinaria, 44
ash, 72; see also Fraxinus
Asparagus, 90
Aster, 32

Barberry, 86, 87
barley, 81

basswood, 41, 77
beech, 41, 103; see also Fagus
bean, 32, 73, 90
beets, 75, 90
Berberis, 86
Betula, 38
birch, 90, 106
blackberry, 73, 92
Boletus, 25
bramble, 18; see also Rubus.

Cabbage, 149
Campanula, 90
Carex, 83
carnation, 69, 85, 90
Carpinus, 94
Cassandra, 97
Catalpa, 69

INDEX TO HOST PLANTS

celery, 69
Chamaecyparis, 91
cherry, 37, 38, 41, 45, 75, 96
chestnut, 73, 107
Chrysopogon, 83
clover, 59, 76, 90
clover weevil, 28
cockle-bur, 42
Comandra, 90
COMPOSITAE, 32, 33, 90
CONVOLVULACEAE, 32
Convolvulus, 83
corn, 81, 83, 90
Cornus, 73, 75
cotton, 73, 78
cowpeas, 78
cranberries, 97
Croton, 90
CRUCIFERAE, 30, 32, 149
cucumber, 73
CUCURBITACEAE, 33
Cupressus, 18
currant, 69, 73

Dandelion, 42
desmids, 23
Dianthus, 83
diatoms, 23

Elm, 41, 72, 125, 129
ERICACEAE, 18, 90, 97
Erigeron, 32
ERYSIBACEAE, 69
Euphorbia, 90

Fagus, 94; see also beech.
Falcata, 23
ferns, 89
fig, 75, 92
fish, 29
flies, 28, 29
Fraxinus, 94; see also ash

Gaylussacia, 97; see also huckleberry
geranium, 32
Glyceria, 83, 85
gooseberry, 41
grape, 22, 30, 31, 41
grass, 32, 41, 43, 44, 81, 83, 84
grasshoppers, 28

hackberry, 41
haw, 41
hawthorn, 18
hazel, 47
hemlock, 95
Hepatica, 84
Hicoria, 94, 97
holly, 53
hollyhock, 73, 90
hop, 41
horseradish, 69
huckleberry, 41; see also Gaylussacia

Impatiens, 33
insects, 44

Juglans, 97
Juncus, 83
Juniperus, 18, 87, 91

Lactarius, 44, 116
LEGUMINOSAE, 18, 92
Lemna, 84
lettuce, 33, 69
lichens, 60, 78
lilac, 41
Linum, 90

Magnolia, 77
mandrake, 88
maple, 41, 53, 69
mayapple, 88
melon, 73
Menispermum, 84
Myosotis, 84

NYCTAGINACEAE, 32

Oak, 41, 97, 107; see also Quercus
oats, 81, 83, 85
Oenothera, 23
onion, 84
Ostrya, 78, 101
Oxalis, 82, 83

Peach, 37, 73, 75, 90
pears, 72, 75
pepper, 73
persimmon, 73
Physalis, 84

pine, 18, 90
pinks, 82
plum, 37, 38, 41, 45, 75, 90
Podophyllum, 88
poplar, 90
Populus, 38
Potamogeton, 85
potato, 22, 30, 31, 32, 75
Potentilla, 18, 38, 92
protozoans, 23
purslane, 32
Pyrola, 90

Quack grass, 84
Quercus, 38, 42, 90 ; see also oak
quince, 72, 85

RANUNCULACEAE, 33
Ranunculus, 84
raspberry, 69, 73, 92
Rhus, 38
Ribes, 90
Rosa, 92
ROSACEAE, 92
rose, 18, 69, 73, 85
Rubus, 91, 92
Russula, 45
rye, 43, 84

Sagittaria, 85

salmon, 29
salsify, 32
Sambucus, 94
Saponaria, 83
Sedum, 89
Sempervivum, 89
shepherd's purse, 30
Solidago, 32
spinach, 73, 75
Sporobolus, 74
spruce, 18
sunflower, 90
sweet clover, 53
sweet potato, 32, 69
thistle, 32
tomato, 73, 75
Trientalis, 84
truffles, 51
turnip, 149

Uredinales, 70, 78

Vaccinium, 90, 97
violets, 69
Virginia creeper, 41

Watermelon, 73, 78
wheat, 41, 81, 83, 84, 85
willow, 41, 53, 90, 106.

III. INDEX OF AUTHORS AND COLLECTORS

Allescher, 72
Anderson, 182
Andrews, 179
Arthur, 172, 173, 174, 180, 186
Arthur & Holway, 93, 180
Atkinson, 30, 39, 79, 121, 145, 153, 165, 166, 186, 188

Banning, 178, 179
Baker, 169
Bay, 37
Bartholomew, 175
Beaumont, 161, 165
Bennett, 191, 192
Berkeley, 161, 197, 198, 199
Berkeley & Curtis, 161, 199

Berlese, 50
Bessey, 169, 173, 174, 183
Blake, 178
Blasdale, 167
Bolles, 178
Bolley, 172, 188
Bose, 160
Brandegee, 198
Brefeld, 27, 29, 85, 94, 96, 97, 132, 159
Brendel, 172
Bresadola, Hennings, & Magnus, 199
Britton, 184
Bulliard, 131, 156
Bundy, 195, 196

Burnap, 141
Burrill, 170, 172
Burrill & Earle, 43, 171, 172
Burrill [& Seymour], 93, 171, 172,
Burt, 66, 134, 194, 202

Calkins, 170
Carpenter, 190
Carleton, 175, 208
Chatin, 52
Chester, 170
Cheney, 196
Clements, 183
Clinton, G. P. 172
Clinton, G. W. 186
Cobb, 179
Cockerell, 169, 186, 199
Collins, 107
Commons, 170
Cook, M. T. 137
Cook, O. F. 186
Cooke, 62, 66, 144, 153, 162, 163, 167, 171, 181, 192, 193, 200, 208
Cooke & Ellis, 185
Cooke & Harkness, 167
Corda, 78, 156
Cragin, 175
Curtis, M. A. 161, 188
Cusick, 190

Davis, 196
Dearness, 97, 158, 207
De Bary, 153, 157, 207
Demetrio, 181
Detmers, 189
De Toni, 143
Dietel, 85, 93, 167, 168
Duby, 53
Dudley, 186
Duggar, 165, 186
Durand, 186

Earle, 165, 166, 169, 186
Ellis, 162, 163, 174, 177, 184, 185, 186, 191, 193
Ellis & Anderson, 182
Ellis & Bartholomew, 176
Ellis & Dearness, 197
Ellis & Everhart, 40, 43, 45, 46, 50, 53, 73, 78, 163, 168, 169, 176, 177, 182, 195, 197, 198

Ellis & Galloway, 182
Ellis & Halsted, 174
Ellis & Holway, 174
Ellis & Kellerman, 176, 190
Ellis & Kelsey, 182, 200
Ellis & Langlois, 177
Ellis & Martin, 171
Everhart, 191. See also Ellis & Everhart.

Fairman, 186, 187
Falconer, 110
Farlow, 24, 33, 46, 84, 93, 167, 168, 179, 184
Farlow & Seymour, 206
Featherman, 177
Fischer, A. 24, 27, 29, 33, 40
Fischer, E. 52, 134
Fisher, 172
Frank, 159
Fries, 104, 131, 144, 157, 198
Frost, 180, 194

Galloway, 181
Gentry, 191
Gerard, 134, 135, 184, 185, 186, 187
Gibson, 144
Gillet, 62, 63, 131
Golden, 37
Griffiths, 69, 182, 192, 196

Haines, 191
Hale, 177
Halsted, 173, 174, 184, 185
Harkness, H. W. 167, 168, 184
Harkness, S. J. 193.
Harkness & Moore, 168
Harper, 43
Harvey, 166, 178
Harvey & Knight, 178
Hay, 145
Heller, 199
Hennings, 96, 97, 107, 114, 116, 119, 122, 124, 131, 168, 200
Hesse, 136
Hicks, 180
Hitchcock, A. S. 173, 175, 200
Hitchcock, E. 179
Holway, 167, 168, 173, 198
Howe, E. C. 186

INDEX OF AUTHORS

Howe, M. A. 167
Humphrey, 29

James, 189
Jennings, 193
Johnson, 180
Jones, M. E. 186, 193
Jones, L. R. 194
Jones & Orton, 194

Kellerman, 175, 176, 189, 190
Kellerman & Swingle, 175, 176
Kellerman & Werner, 190
Kelsey, 182

Langlois, 177, 178
Lapham, 161, 195
Lea, 189, 190
Leville, 43
Linnaeus, 155
Lindau, 43, 45, 46, 50, 53, 54, 62, 94, 96
Lister, 153
Lloyd, C. G. 119, 189
Lloyd, F. E. 190
Lodeman, 21
Ludwig, 159

Macadam, 116
Macbride, 153, 170, 173, 174, 198
Macbride & Allin, 173
Macbride & Hitchcock, 175
Macbride & Smith, 198
MacMillan, 180
Macoun, 170, 191, 197
Martin, 40, 43, 72, 153, 160, 200, 206
Massee, 45, 66, 116, 131, 141
McCarthy, 188
McClatchie, 167, 168
Michael, 145
Micheli, 155
Michener, 161
Migula, 154
Millspaugh & Nuttall, 195
Montagne, 200
Morgan, 79, 102, 116, 118, 119, 120, 122, 123, 124, 127, 128, 131, 134, 137, 141, 153, 177, 189, 190
Mühlenberg, 191, 192

Nash, 170
Nelson, 196
Norton, 175, 176
Nuttall, L. W. 195

Olive, 172, 173
Olney, 161, 191

Pammel, 172, 183
Parker, 195
Pasteur, 37
Patterson, 39
Peck, 100, 115, 116, 117, 118, 119, 121, 122, 123, 124, 125, 126, 127, 128, 131, 141, 144, 162, 166, 168, 170, 179, 186, 187
Persoon, 156
Peters, 161, 165
Pettit, 79
Phillips, 63, 168
Phillips & Harkness, 169
Piper, 195
Plowright, 85, 93, 169
Plowright & Harkness, 169
Porter & Coulter, 170
Pound, 28, 183
Pound & Clements, 79
Purpus, 167, 168

Rafinesque, 160
Ravenel, 161, 162, 170, 171, 192, 193
Rehm, 53, 54, 62, 66
Rex, 184, 191
Richards, 97
Robinson, B. L. 39
Robinson, W. 110
Rolfs, 170
Rose, 172
Rostafinski, 153
Rostrup, 197, 200, 208
Roussel, 200
Rusby, 186
Rydberg, 182

Saccardo, 14, 15, 24, 27, 28, 29, 33, 37, 38, 40, 43, 45, 46, 50, 52, 53, 54, 62, 66, 72, 73, 78, 85, 93, 94, 95, 96, 97, 105, 107, 131, 134, 136, 142, 143, 158.
Sadebeck, 39

Sagra, 200
Saunders, 183
Schaeffer, 131, 156
Schroeter, 24, 27, 28, 29, 33, 37, 39, 66, 153
Schweinitz, 160, 188, 191
Selby, 189, 190
Setchell, 85, 170
Seymour, 171, 182, 183, 188, 196
Seymour & Earle, 159
Shear, 183, 186
Smith, C. L., 199
Smith, E. F., 207
Smith, J. G., 183
Smyth, 176
Snyder, 172, 173
Soraurer, 159
Sowerby, 131, 156
Spalding, 180
Sprague, 161, 180
Stevens, 184, 185, 189, 190
Stevenson, J. 206.
Stevenson, W. C. 184, 191
Stewart, 186
Stoneman, 75
Sturgis, 21, 170
Suksdorf, 195
Sullivant, 189
Swartz, 199, 200
Swingle, 33, 170, 175, 176

Thaxter, 29, 51, 79, 93, 134, 152
Thümen, de, 192, 197
Tournefort, 155

Tracy, 169, 181, 186
Tracy & Earle, 181
Trelease, 141, 196
Tubeuf & Smith, 159
Tuckerman & Frost, 180
Tulasne, 43, 45, 50, 94, 96, 97, 136, 142, 157, 158

Underwood, 66, 166, 170, 172, 173, 186
Underwood & Earle, 90, 91, 166

Vittadini, 144
Van Tieghem, 27, 28
Van Tieghem & Monnier, 27
Vize, 169
Von Tafel, 159

Waghorne, 197
Waite, 172
Walters, 177
Webber, 170, 183, 184
Webster, 141
Wheeler, 180
Williams, 182, 183, 192, 196
Wingate, 191
Winter, 46, 85, 93, 181
Winter & Demetrio, 181
Winter & Rehm, 50
Woods, 183
Wooton, 186
Wright, C. 161, 170, 199

Zopf, 16, 37, 159

IV. General Index of Subjects and Explanation of Technical Terms

Acervuli, tufts of mycelium bearing spores
adnate, squarely attached to the stem (*Pl. 7. f. 1*)
aecidiospores, the spores produced in cluster-cups
aethallium, 148
Alabama, exploration for fungi in, 165
Alaska, exploration for fungi in, 166
allantoid, curved like a crescent with rounded ends (*Pl. 1. f. 6*)
anastomosing, uniting together to form a network
animals distinguished from green plants, 5, 6; compared with green plants, 9, 10; compared with fungi, 7

annulus, 110; the ring formed on the stem in certain mushrooms by the
  veil separating from the margin of the pileus
antherid, the male reproductive apparatus in the lower plants
anthracnose, 73; a disease caused by parasitic species of Melanconiales
apothecium, 34; the ascoma of lichens
arachnoid, cobwebby
Arizona, exploration for fungi in, 166
Arkansas, exploration for fungi in, 166
asci, 14, 18, 34; membranous sacs containing spores (*Pl. 1. f. 17*)
ascocarp, 34; a collective term for the body containing asci
ascoma, 34; the disc-like body bearing the asci in the Pezizales and their
  allies

Bacteria, 154
basidia, 14, 18; enlarged cells bearing spores in the Basidiomycetes (*Pl. 1.
  f. 18, 19*)
bird's nest fungi, 81, 141, 142
black knot, 45; a disease of cherry and plum caused by *Plowrightia*
bladder plums, 38
Bordeaux mixture, 20
bracket fungi, 80, 99
Brefeld, contribution to mycology by, 159
bricktops, 123

California, Exploration for fungi in, 167
campanulate, bell-shaped
Canada, exploration for fungi in, 197
capillitium, simple or branched threads mixed with spores
cedar-apple, 87, 91
Central America, exploration for fungi in, 198
chanterelle, 115
chlamydospores, 82; the reproductive bodies of smut
circumscissile, breaking apart along the equatorial line
club-foot of cabbage, 149; a disease caused by *Plasmodiophora*
cluster-cups, 86; the first stage of many rusts
collecting fungi, methods of, 201–203
Colorado, explorations for fungi in, 169
columella, the extension of the stalk into the sporangium or peridium
conidia, 34; dust-like spores usually produced directly from the hyphae
  (*Pl. 5. f. 9, 10*)
conidiophores, enlarged ends of hyphae bearing conidia
conjugation, 14; reproduction by equal gametes
Connecticut, exploration for fungi in, 170
copper acetate as a fungicide, 20
copper carbonate as a fungicide, 20
copper sulphate as a fungicide, 20
coral-fungi, 101, 103
corrosive sublimate as a fungicide, 21
cup-fungi, 36, 54
Curtis, contribution to American mycology by, 161

Damping off, 29
De Bary, contribution to mycology by, 158
decurrent, extending down the stem (*Pl. 7. f. 4, 6*)
Delaware, exploration for fungi in, 170
determination of species, 206
devil's snuftboxes, 136
diatoms, 9
dictyoid, the same as muriform, *q. v.*
didymoid, twin; composed of two cells (*Pl. 1. f. 7-9*)
Discomycetes, 35; an obsolete group name for the Pezizales and their allies
downy mildews, 29; a group of fungi belonging to the Peronosporales

Earth-stars, 137, 140
eccentric, attached at one side of the centre
edible fungi, 144
Ellis, contribution to American mycology by, 162, 163
endospore, the inner wall of a spore.
epixylous, growing on wood
ergot, 43, 44; a disease of rye caused by *Claviceps*
erumpent, breaking through the bark or epidermis

Fairy-ring, 118
ferns, 8, 9
field mushroom, 122
field notes, 203, 204
fission in fungi, 13
Florida, exploration for fungi in, 170
fly-agaric, 119, 120
Fries, contribution to mycology by, 157
fungi, chemistry of, 15, 16; conditions of growth, 16; classes of, 18, 19; species among, 19; germination of, 20; origin of, 12; relation to algae, 11, 12, 67; distinguished from green plants, 6, 7; compared with animals, 7; number of, 10; reproduction in, 13
fungicides, 20, 21
Fungi exsiccati, of Arthur and Holway, 93; of Ellis, 163; of Kellerman and Swingle, 175; of Ravenel, 161; of Seymour and Earle, 159; of Shear, 186
fungi imperfecti, 35, 68
fungous diseases, 20, 21, 22
fungus cellulose, 13

Gemmation, 13, 36; reproduction by budding
Georgia, exploration for fungi in, 171
gleba, the gelatinous spore-mass in the Phallales
grape mildew, 30, 31
Greenland, exploration for fungi in, 197
green plants distinguished from animals, 5, 6; from fungi, 6, 7; compared with animals, 9, 10

Haustorium, 13; a projecting portion of a hypha which penetrates a cell of a host enabling the fungus to obtain its supply of food; sometimes the hypha merely forms a disc on the surface of the cell

heteroecism, the habit of living on more than one host-plant during different periods of the life history of a parasitic species
host, a plant or animal supporting a parasite
host index of fungi, 206 (see also Index II)
hot water as a fungicide, 21
hygrophanous, watery in appearance as tho saturated
hymenium, the membrane in which the basidia of mushrooms and their allies are borne
hypha, 13; the thread-like vegetative part of a fungus
Hyphomycetes, 74; a collective term for the Moniliales; in certain recent text-books improperly used for all fungi
hypothallus, a membranous or fleshy base to which perithecia or sporangia are attached
hypothecium, 54; the upper stratum of the ascoma containing the asci
hysterioid, elongate boat-shaped like one of the Hysteriaceae

Idaho, Exploration for fungi in, 171
Illinois, exploration for fungi in, 171
Indiana, exploration for fungi in, 172
infundibuliform, funnel-shaped
ink-caps, 115; species of *Coprinus*
intermediate forms of fungi, 130
Iowa, exploration for fungi in, 173

Jew's ear, 94; species of *Auricularia*

Kansas, Exploration for fungi in, 175
Kentucky, exploration for fungi in, 177

Leaf-blight, 74
leaf-spot, 69
lenticular, lens-shaped
lichens, 67
Linnaeus, contribution to mycology by, 155
Louisiana, exploration for fungi in, 177
liverworts, 9

Maine, Exploration for fungi in, 178
Maryland, exploration for fungi in, 178
Massachusetts, exploration for fungi in, 179
Mexico, exploration for fungi in, 198
Micheli, contribution to mycology by, 155
Michigan, exploration for fungi in, 180
mildews, downy, 29; powdery, 40
milk-fungi, 117, 118
Minnesota, exploration for fungi in, 180
Mississippi, exploration for fungi in, 181
Missouri, exploration for fungi in, 181
Montana, exploration for fungi in, 182
morcheln, 66
morel, 36, 65, 66, 144 (see also frontispiece)

mosses, 8, 9
moulds, 24, 25, 26
muriform, with septa extending in more than one plane so as to give the appearance in optical section of a wall of masonry (*Pl. 1. f. 15*)
mushrooms, 99, 109; cultivation of, 110
mycelium, 13; a collective term for hyphæ growing in interlacing masses
mycological collection at the New York Botanical Garden, 163; at Harvard University, 161; at New York State Museum, 162; at Department of Agriculture, Washington, 170; at Alabama Polytechnic Institute, 166; at the Missouri Botanic Garden, 182; at California Academy of Sciences, 167; at Philadelphia Academy of Science, 160; at University of Nebraska, 183; at Cincinnati (C. G. Lloyd's), 189
mycology, history of, 155-159; in America, 160-164

Nebraska, Exploration for fungi in, 183
Nevada, exploration for fungi in, 184
New Hampshire, exploration for fungi in, 184
New Jersey, exploration for fungi in, 184
New Mexico, exploration for fungi in, 186
New York, exploration for fungi in, 186
North Carolina, exploration for fungi in, 188
North Dakota, exploration for fungi in, 188

Ohio, Exploration for fungi in, 189
Oklahoma, exploration for fungi in, 189
oöspore, 14; the resting spore resulting from the fertilization of an egg by an antherid or sperm cell
Oregon, exploration for fungi in, 189
oyster mushrooms, 124, 125

Paraphyses, 34; sterile, simple or branched bodies interspersed with the asci
parasite, an organism living at the expense of another
parasol-fungi, 121; species of *Lepiota*
parenchymatous, formed of thin-walled cells of nearly equal diameter in every direction
patelliform, having the shape of the knee-cap (patella)
peach-curl, 31, 38
Peck, contribution to American mycology by, 162
Pennsylvania, exploration for fungi in, 191
peridium, 54; a more or less thickened covering to a puff-ball
perithecium, 34; a rounded, oval, pyriform or beaked structure in which the asci are developed (*Pl. 4, f. 5, 16, 17*)
Persoon, contribution to mycology by, 156, 157
phragmoid, divided by two or more septa at right angles to the long axis
pileate, with a cap or pileus like that of a mushroom or toadstool
plant diseases, bureau of, 21
plasmodiocarp, 148
plasmodium, 146
poison-cup, 119 (*Pl. 8*)
poisonous fungi, 119, 120
pond-scums, 8, 9

potassium sulphide as a fungicide, 21
potato rot, 31
powdery mildews, 18, 40; parasitic fungi belonging to the Perisporiales
preservation of fungi, methods of, 202-206
promycelium, the early hypha produced by a germinating spore
pseudoperidium, a membraneous covering to certain aecidial spores
puff-balls, 81, 136, 137, 141; subterranean, 135!; thick-skinned, 143
pycnidia, 68; perithecia-like cavities producing spores from the inner surface of their walls
Pyrenomycetes, 36; an obsolete group name for the Sphaeriales and their allies
pyriform, pear-shaped

Receptacle, 34
respiration in plants and animals, 9
resupinate, without a pileus
Rhode Island, exploration for fungi in, 191
ripe rot, 73
rusts, 80; parasitic fungi belonging to the Uredinales

Saccardo, Contribution to mycology by, 158
salmon disease, 22, 29
saprophyte, a fungus growing on dead organic matter
scab, 75
Schweinitz, contribution to American mycology by, 160
sclerotia, 43; hard bodies which serve as receptacles of reserve food material
sea lettuce, 9
sea weeds, 8, 9
seed plants, 8, 9
septa, 13; partitions extending across the tube of a hypha or a spore
sinuate, with a notch near the junction with the stem (*Pl. 7. f. 2*)
slime-moulds, 146, 149; relation to protozoans, 152; relation to bacteria, 152; relation to moulds, 152
smoke-balls, 136
smut in grain, 20, 21
smuts, 80; parasitic fungi belonging to the Ustilaginale
sorus, a mass of spores breaking through the epidermis of a host plant
South Carolina, exploration for fungi in, 192
South Dakota, exploration for fungi in, 192
spermatia, 88; spore-like bodies produced from spermogonia
spermogones, the same as spermogonia, *q. v.*
spermogonia, 88; a form of reproduction among the Uredinales whose relations are imperfectly known
sporangia, 147; spore cases
sporangioles; spore receptacles with persistent walls, found in the Nidulariales and their allies
spores, color of, 15; shape of, 15
spore-prints, 111
spring mushroom, 144 (see also morel)
sterigmata, 14; slender projections from the basidia which bear spores

stinkhorns, 81, 132, 133, 134
stroma, a compact substance formed of mycelium uniting the perithecia, or in which the perithecia are imbedded
sweetbread-mushroom, 126

Teleutospore, the winter or resting spore of the rusts (Uredinales)
Tennessee, exploration for fungi in, 192
terminology of groups, 19, 20
Texas, exploration for fungi in, 193
toadstools, 99
truffles, 35
Tulasne, contribution to mycology by, 158

Umbilicate, provided with a depression like the navel (*Pl. 7. f. 5*)
umbonate, provided with a small raised prominence (*Pl. 7. f. 4*)
uredospores, the thin-walled summer spores of a rust
Utah, exploration for fungi in, 193

Vermont, Exploration for fungi in, 194
Virginia, exploration for fungi in, 194
 volva, 110; a covering or universal veil of certain mushrooms which in the expanded form appears either as floccose scales on the pileus, or as a cup at the base of the stem, or both (*cf. Pl. 8*)

Washington, Exploration for fungi in, 195
West Indies, exploration for fungi in, 199
West Virginia, exploration for fungi in, 195
wilt, 78; a disease caused by species of *Fusarium*
Wisconsin, exploration for fungi in, 195
Wyoming, exploration for fungi in, 196

Yeast, 35

Zoöspore, a reproductive body provided with cilia and capable of motion
zygospore, 14; the resting spore resulting from the union of two like gametes or sexual cells (*Pl. 2, f. 3, 4*)

EXPLANATION OF PLATE

Forms of Spores

FIG. 1. Spherical spore.
FIG. 2. Oval hyaline spore.
FIG. 3. Oval biguttulate spore.
FIG. 4. Oval biguttulate muricate spore.
FIG. 5. Lemon-shaped guttulate spore.
FIG. 6. Allantoid spore.
FIG. 7. Didymoid (2-celled) spore; the cells unequal.
FIG. 8. Didymoid (2-celled) symmetrical spore.
FIG. 9. Didymoid (2-celled) appendaged spore.
FIGS. 10, 11, 12. Various types of phragmoid spores.
FIG. 13. Filiform septate spore.
FIG. 14. Filiform non-septate spore.
FIG. 15. Dictyoid (muriform) spore.
FIG. 16. Spore of *Peziza aurantia* showing surface reticulations.
FIG. 17. Ascus containing spores.
FIG. 18. Basidium with sterigmata.
FIG. 19. Basidiospores on the sterigmata.

All the figures are greatly magnified and were redrawn for the most part from various sources.

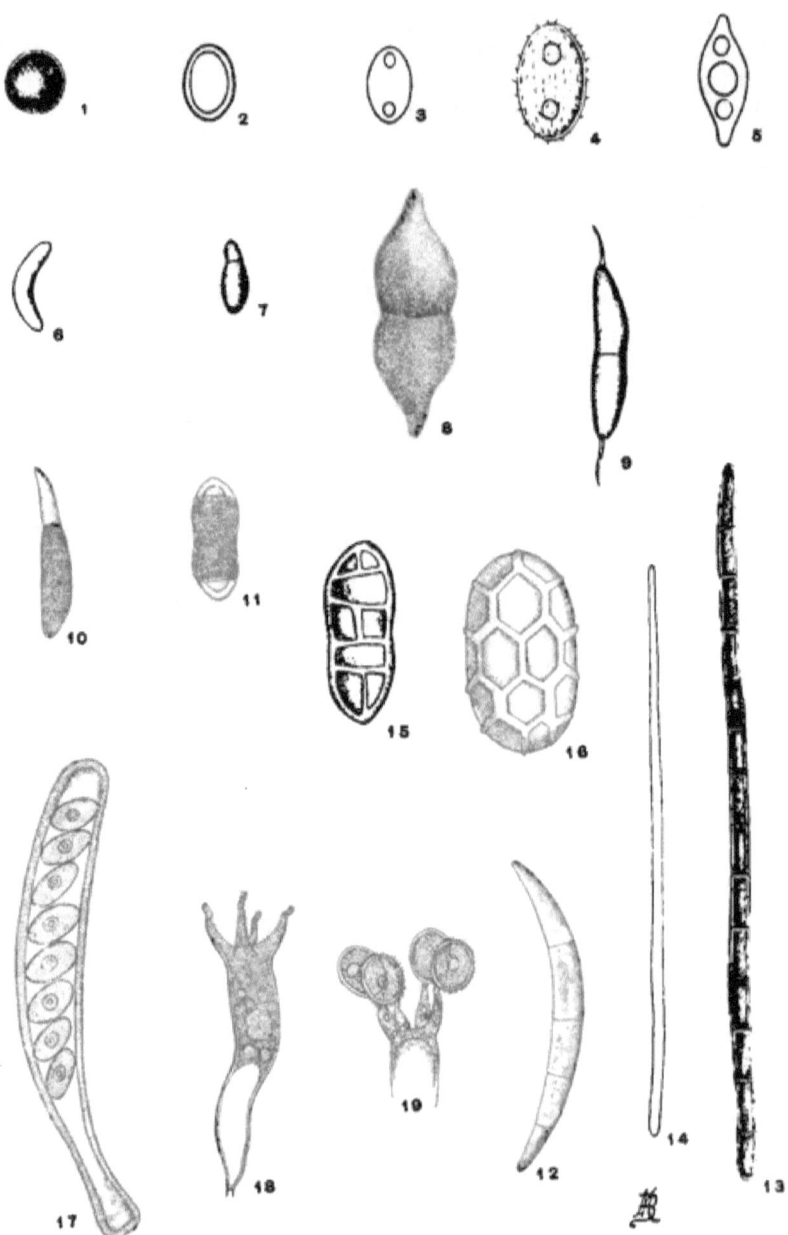

EXPLANATION OF PLATE II

PHYCOMYCETES

FIG. 1. *Septocarpus corynephorus* (CHYTRIDIALES) parasitic on diatom. X 500. (Redrawn from Zopf.)

FIG. 2. *Mucor Mucedo* (MUCORALES) showing root-like hyphae and sporangia borne on aerial hyphae. X 17. (Redrawn from Kerner.)

FIG. 3. *Mucor stolonifer* (MUCORALES). Early stages of conjugation: A. Two hyphal branches approaching; B. The same still farther advanced; C. Suspensors cut apart from the gametes by septa. X 60. (Redrawn from De Bary.)

FIG. 4. *Mucor Mucedo* (MUCORALES). Mature zygospore showing the suspensors above and below. X 100. (Redrawn from Brefeld.)

FIG. 5. *Piptocephalis Freseniana* (MUCORALES). Conidia X 200. (Redrawn from Brefeld.)

FIG. 6. *Piptocephalis Freseniana* (MUCORALES). Haustoria attached to the hypha of *Mucor* (shown by the two parallel lines at the left). X 400. (Redrawn from Brefeld.)

FIG. 7. *Piptocephalis Freseniana* (MUCORALES). Sexual reproduction, showing enlarged suspensors and spinulose zygospore. X 300. (Redrawn from Brefeld.)

FIG. 8. *Pilobolus Kleinii* (MUCORALES). Sporangial stage. X 150. (Redrawn from Brefeld.)

FIG. 9. *Pilobolus crystallinus* (MUCORALES). Sexual reproduction showing suspensors and zygospore. X 65. (Redrawn from Zopf.)

EXPLANATION OF PLATE III

PHYCOMYCETES

FIG. 1. *Synchytrium mercurialis* (CHYTRIDIALES). Producing a large gall on the epidermis of *Mercurialis perennis*. The resting spore appears in the large central cell of the host. X 90. (Redrawn from Woronin.)

FIG. 2. *Synchytrium myosotidis* (CHYTRIDIALES). Producing a smaller gall in a single epidermal cell of *Myosotis stricta*, which encloses an oval resting spore. X 110. (Redrawn from Schroeter.)

FIG. 3. *Plasmopara viticola* (PERONOSPORALES). Haustoria penetrating a cell of the host; the hypha of the fungus appears on the lower side of the cell and is shaded.

FIG. 4. *Peronospora* (PERONOSPORALES). Conidiophore bearing solitary conidia issuing from a stoma on the under surface of a leaf; FIG. 5. Sexual reproduction showing antherid fertilizing the egg.

FIG. 6. *Albugo candida* (PERONOSPORALES). Conidia borne in chain-like rows. X 220. (Redrawn from De Bary.)

FIG. 7. *Empusa* (ENTOMOPHTHORALES) on the larva of clover-weevil (*Phytonomus*) which has crawled to the tip of a blade of grass to die. X 5. (Redrawn from Arthur.)

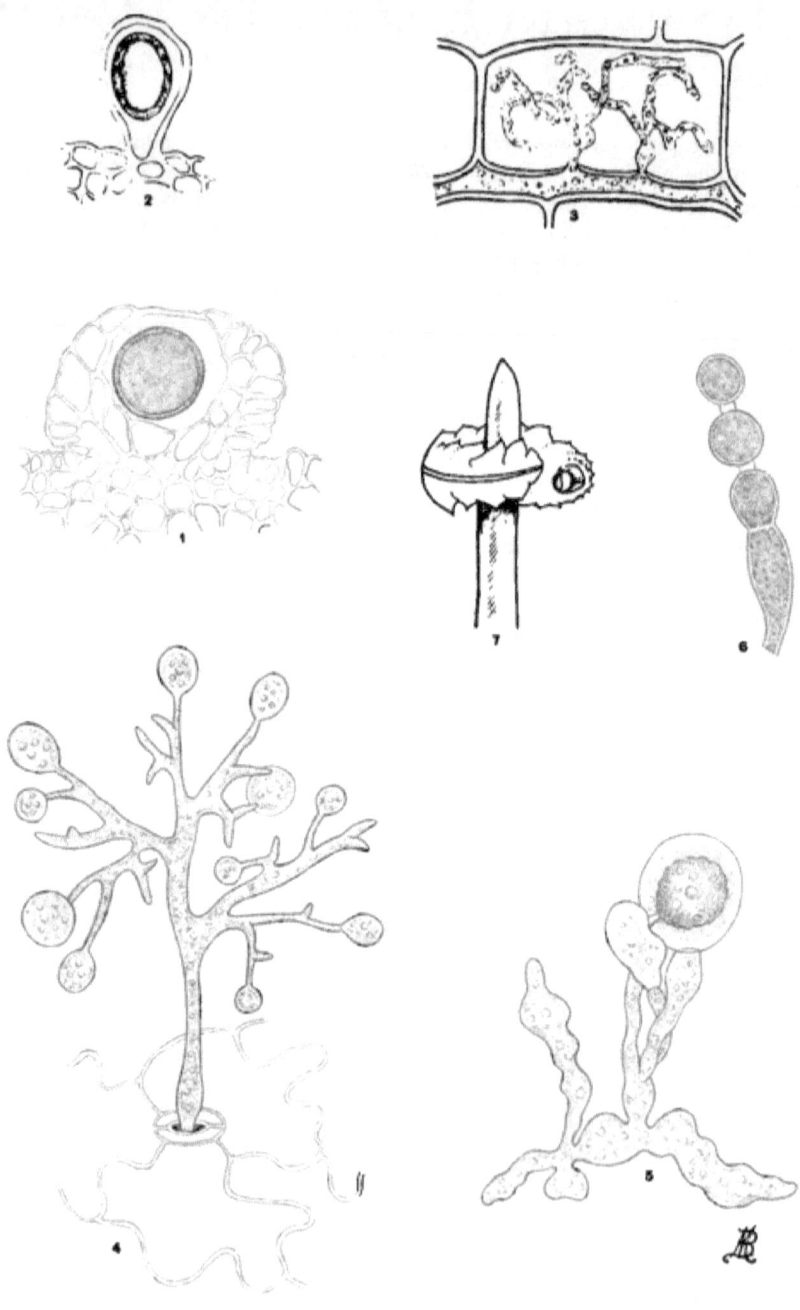

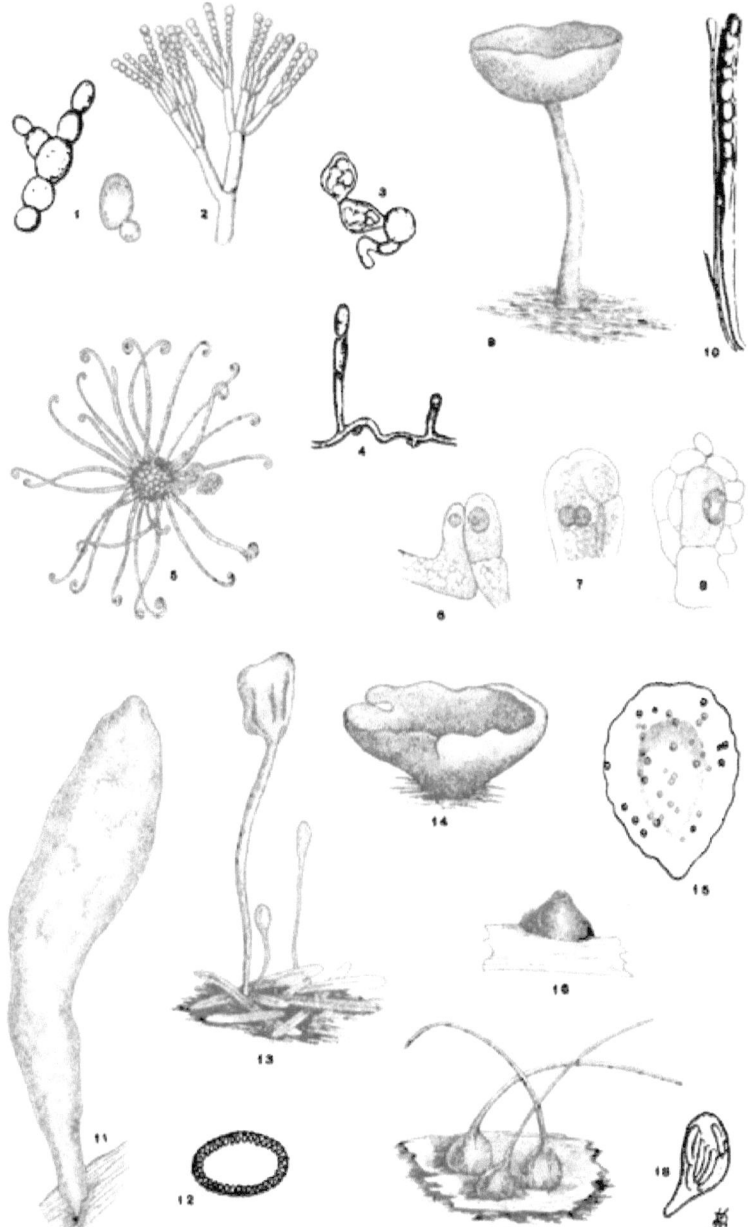

EXPLANATION OF PLATE IV

ASCOMYCETES

FIG. 1. Yeast cells of *Saccharomyces cerevisiae* (SACCHAROMYCETALES) normally budding. × 550. (Redrawn from Reess.)

FIG. 2. *Penicillium crustaceum* (ASPERGILLALES). Conidial stage seen in ordinary green mould. × 160. (Redrawn from Brefeld.)

FIG. 3. *Penicillium crustaceum* (ASPERGILLALES). Ascosporic stage, the asci produced in skeins. × 500. (Redrawn from Brefeld.)

FIG. 4. *Uncinula necator* (PERISPORIALES). Conidia from the powdery mildew of the grape. × 130. (Redrawn from Scribner.)

FIG. 5. *Uncinula necator* (PERISPORIALES). Perithecium showing hooked appendages and asci protruding from the crushed body. × 40. (Redrawn from Scribner.)

FIGS. 6, 7, 8. Successive stages in the sexual reproduction of *Sphaerotheca Castagnei* (PERISPORIALES). Greatly magnified. (Redrawn from Harper.)

FIG. 9. *Peziza macropus* (PEZIZALES). Natural size. (Redrawn from Lindau.); FIG. 10. Paraphyses, asci, and spores. × 200. (Redrawn from Rehm.)

FIG. 11. *Xylaria polymorpha* (SPHAERIALES), showing habit; Fig. 12, section across the stroma showing perithecia; both natural size.

FIG. 13. *Mitrula phalloides* (HELVELLALES); natural size.

FIG. 14. *Peziza aurantia* (PEZIZALES); natural size.

FIG. 15. Inside of half a peach pit showing perithecia of *Caryospora putaminum* (SPHAERIALES); natural size.

FIG. 16. *Caryospora putaminum* (SPHAERIALES); side view of perithecium showing the ostiolum. × 15. (Redrawn from Winter.)

FIG. 17. *Ceratostomella pilifera* (SPHAERIALES). Three perithecia on a fragment of wood, showing the long beak-like ostiola. × 30; Fig. 18, ascus containing allantoid spores. Strongly magnified. (Redrawn from Lindau.)

EXPLANATION OF PLATE V

FUNGI IMPERFECTI

FIG. 1. *Septoria pirina* (SPHAEROPSIDALES). Pear leaf showing leaf-spots. The pycnidia appear as minute dots one or more on each spot. Natural size; drawn from life. FIG. 2. Section through a pycnidium on the pear leaf showing the attachment of the spores. Greatly magnified. (Redrawn from Duggar.)

FIG. 3. Pycnidium of *Ampelomyces quisqualis* (SPHAEROPSIDALES), inside the conidiophore of *Erysibe*. X 200. Drawn from nature by D. Griffiths.

FIG. 4. *Entomosporium* (SPHAEROPSIDALES); spores from the leaf-blight of the pear. Greatly magnified. (Redrawn from Duggar.)

FIG. 5. *Dinemosporium* (SPHAEROPSIDALES). Septate appendaged spore. Greatly magnified.

FIG. 6. *Colletotrichum* (MELANCONIALES); bean pod affected with anthracnose caused by the fungus. ½ natural size. (Redrawn from Cowing); Fig. 7, spores of same, greatly magnified. (Redrawn from Southworth.)

FIG. 8. *Pestalozzia* (MELANCONIALES). Spores showing appendages and hyaline end-cells. X 400. (Redrawn from Desmazières.)

FIG. 9. *Monilia fructigena* (MONILIALES), hypha forming catenulate spores. Greatly magnified.

FIG. 10. *Botrytis vulgaris* (MONILIALES). End of spore-bearing hypha with clusters of spores.

FIG. 11. *Ramularia* (MONILIALES). Didymoid spore greatly magnified.

FIG. 12. Acervulus of *Cercospora gossypina* (MONILIALES) issuing from the epidermis of cotton leaf; Fig. 13, spores of same. Both greatly magnified. (Redrawn from Southworth.)

FIG. 14. *Ceratophorum* (MONILIALES). Spore bearing appendages at either end. Greatly magnified.

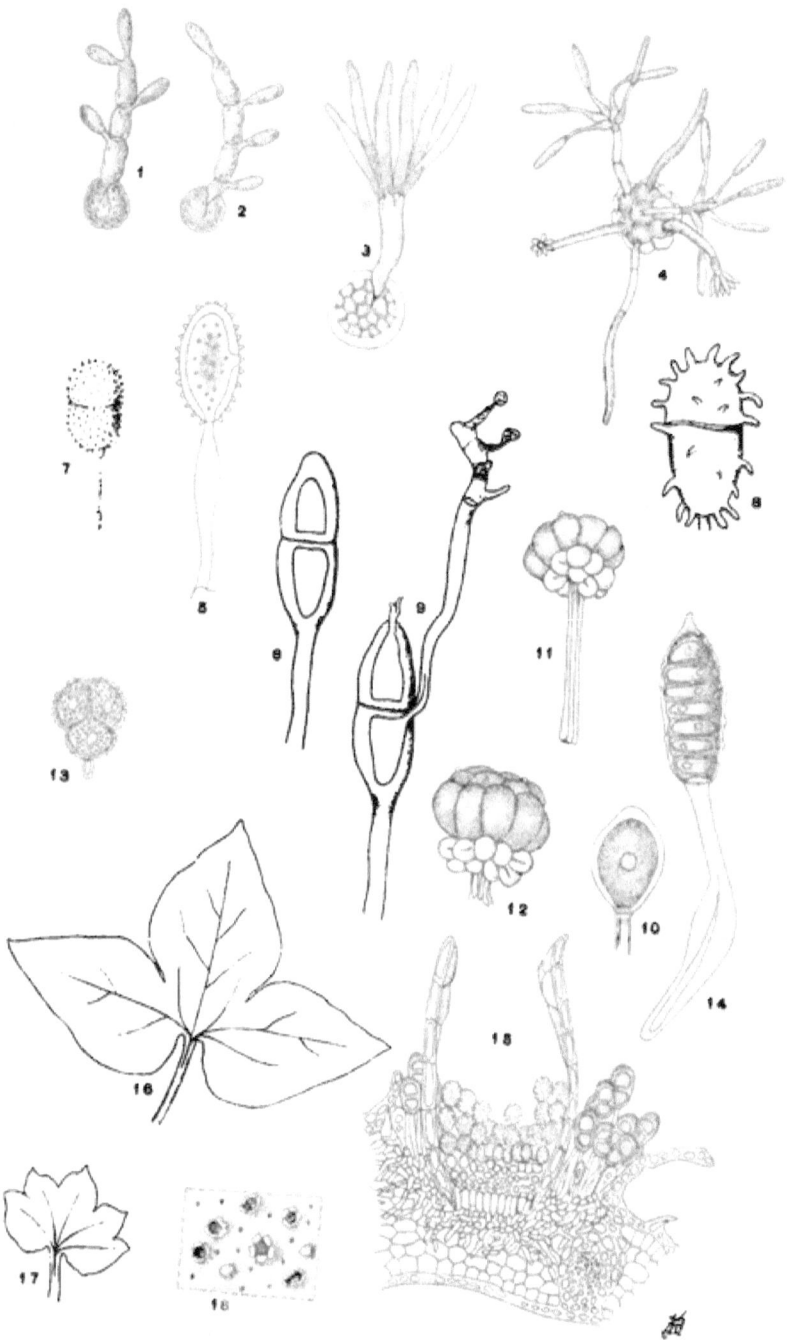

EXPLANATION OF PLATE VI

BASIDIOMYCETES

FIGS. 1, 2. *Ustilago avenae* (USTILAGINALES). Germinating chlamydospores producing spores laterally and terminally. × 350. (Redrawn from Brefeld.)

FIG. 3. *Tilletia zonata* (USTILAGINALES). Germinating chlamydospore producing a cluster of spores at the apex. × 300. (Redrawn from Brefeld.)

FIG. 4. *Urocystis violae* (USTILAGINALES). Germination of a chlamydospore. × 180. (Redrawn from Brefeld.)

FIG. 5. *Puccinia graminis* (UREDINALES). Uredospore with separable pedicel. × 200. (Redrawn from Sachs.)

FIG. 6. *Puccinia graminis* (UREDINALES). Teleutospore. × 330. Redrawn from Peck).

FIG. 7. *Puccinia anemones* (UREDINALES). Teleutospore. × 300. (Redrawn from Peck.)

FIG. 8. *Puccinia podophylli* (UREDINALES). Teleutospore. × 300. (Redrawn from Peck.)

FIG 9. *Puccinia graminis* (UREDINALES). Germinating teleutospore producing the basidiospores from the upper part of the promycelium. × 330. (Redrawn from Sachs.)

FIG. 10. *Uromyces trifolii* (UREDINALES). Teleutospore. × 375. (Redrawn from Dietel.)

FIGS. 11, 12. *Ravenelia cassiicola* (UREDINALES). Compound teleutospores with long and short pedicels respectively. × 200. (Redrawn from Dietel.)

FIG. 13. *Triphragmium ulmariae* (UREDINALES). Teleutospore. × 250. (Redrawn from Dietel.)

FIG. 14. *Phragmidium mucronatum* (UREDINALES). Teleutospore. × 200. (Redrawn from Scribner.)

FIG. 15. *Puccinia graminella* (UREDINALES). Section across a grass leaf showing a cluster cup (with the two cut portions of its pseudoperidium projecting upwards) containing aecidiospores, together with teleutospores (two-celled) rising from the same mycelium. × 150. (Redrawn from Dietel.)

FIG. 16. Normal leaf of *Hepatica acuta*. FIG. 17, leaf from the same plant distorted by *Aecidium hepaticatum* (UREDINALES). Both one-half natural size. FIG. 18, a portion of the under surface of the leaf in Fig. 17 showing the crater-like cluster-cups with spermogonia scattered among them. × 5.

EXPLANATION OF PLATE VII

BASIDIOMYCETES

FIG. 1. Section of an agaric showing solid fleshy stem and adnate lamellae.

FIG. 2. Section of an agaric showing hollow stem and sinuate lamellae.

FIG. 3. Section of an agaric showing hollow stem and free lamellae.

FIG. 4. Section of an agaric showing an umbonate top-shaped pileus, solid fleshy stem, and decurrent lamellae.

FIG. 5. *Clitocybe fragrans* (AGARICALES.) Showing umbilicate pileus, and decurrent lamellae. (Redrawn from Cooke.)

FIG. 6. *Clitocybe cyathiformis* (AGARICALES.) Showing infundibuliform pileus, and decurrent heterophyllous lamellae. (Redrawn from Cooke.)

FIG. 7. *Myriostoma coliforme* (LYCOPERDALES). Natural size. (Redrawn mainly from Morgan.)

Pl. 8.

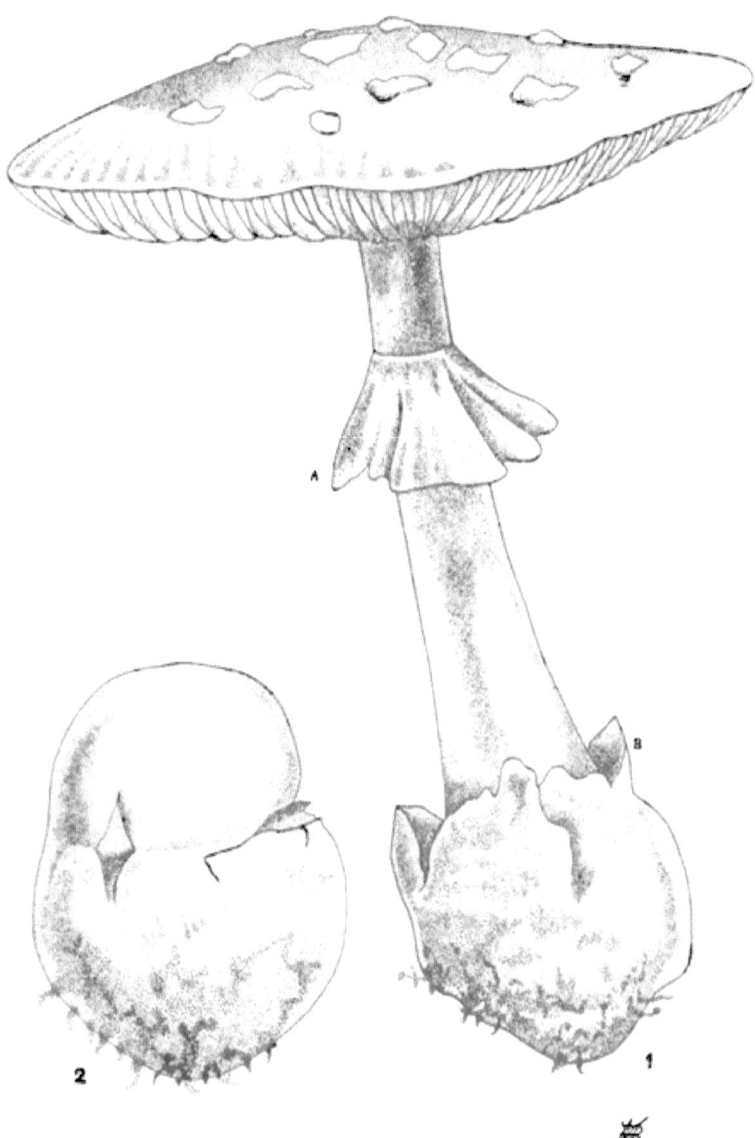

EXPLANATION OF PLATE VIII

BASIDIOMYCETES

FIG. 1. Poison cup, *Amanita phalloides* (AGARICALES). Two-thirds natural size. This is the most dangerous of the species of poisonous fungi. It can be recognized by its white lamellae, its skirt-like annulus (A), and its volva (B) at the base of the stem. A part of the volva was also carried up with the expanding pileus and appears on the surface as a series of floccose separable scales.

FIG. 2. The same in an early condition showing the volva completely investing the pileus which is not yet expanded.

EXPLANATION OF PLATE IX

BASIDIOMYCETES

FIG. 1. *Simblum rubescens* (PHALLALES). Expanded form showing the cup-like peridium at the base, and the latticed receptacle at the summit. One-half natural size. (Redrawn from Gerard.)

FIG. 2. *Phallogaster saccatus* (PHALLALES), just before the rupture of the peridium. FIG. 3. The same ruptured, exposing the gleba (the dark shaded portions) within the peridium. Both somewhat enlarged. FIG. 4. A cluster of basidia, some of them bearing spores. $\times$ 800. (All redrawn from Thaxter.)

FIGS. 5, 6, 7. *Catastoma circumscissum* (LYCOPERDALES). Natural size. Fig. 5 shows the outer peridium rupturing at the equator; Fig. 6 shows the upper part removed with the inner peridium which opens at a point originally at the bottom of the latter with a single crater; Fig. 7 shows the cup left in the ground after the part shown in Fig. 6 became free. (Redrawn from Morgan.)

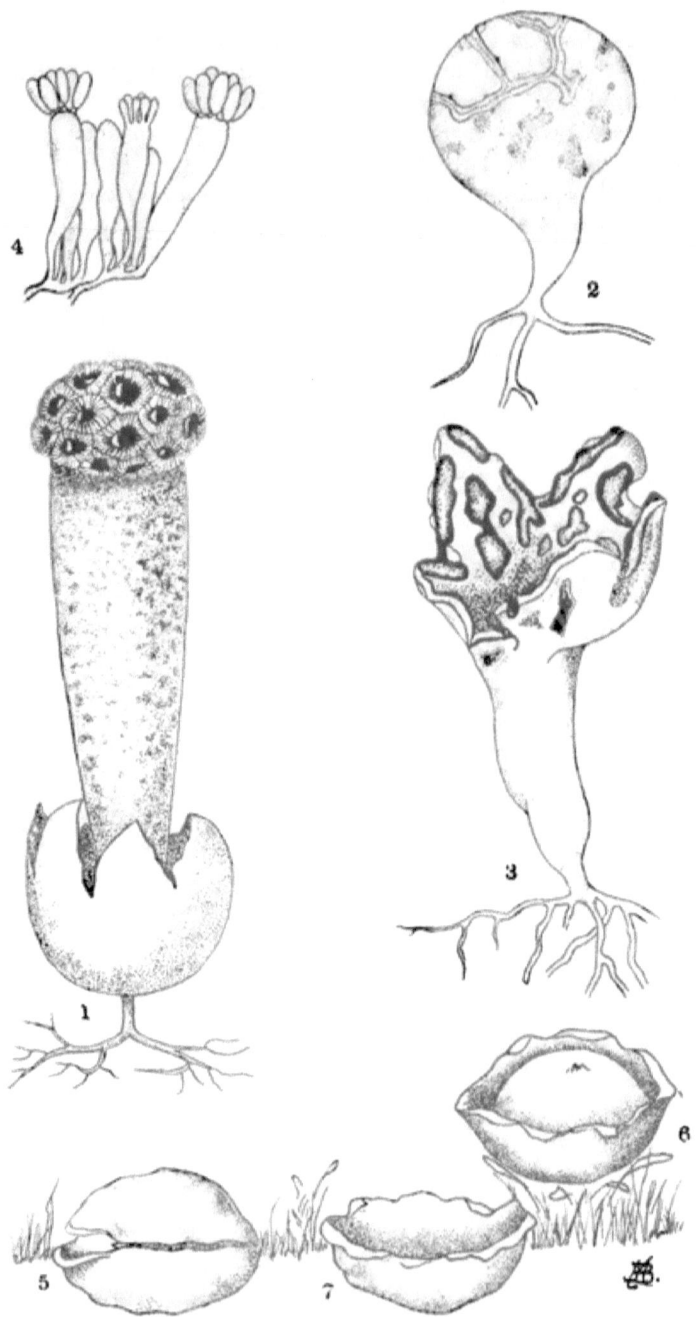

www.ingramcontent.com/pod-product-compliance
Lightning Source LLC
Chambersburg PA
CBHW021345230426
43666CB00006B/409